46 Advances in Polymer Science

Fortschritte der Hochpolymeren-Forschung

AF772169

Behavior of Macromolecules

With Contributions by
R. C. Arridge P. J. Barham M. Kawaguchi
J. Kolařík A. Takahashi

With 60 Figures

Springer-Verlag
Berlin Heidelberg GmbH 1982

Editors

ISBN 978-3-662-15772-5 ISBN 978-3-540-39453-2 (eBook)
DOI 10.1007/978-3-540-39453-2

Library of Congress Catalog Card Number 61-642

Originally published by Springer-Verlag Berlin Heidelberg New York in 1982
Softcover reprint of the hardcover 1st edition 1982

2152/3140 – 543210

Table of Contents

The Structure of Macromolecules Adsorbed on Interfaces
A. Takahashi and M. Kawaguchi 1

Polymer Elasticity – Discrete and Continuum Models
R. C. Arridge and P. J. Barham 67

Secondary Relaxations in Glassy Polymers – Hydrophilic Polymethacrylates
J. Kolařík 119

Author Index Volumes 1–46 163

The Structure of Macromolecules Adsorbed on Interfaces

Akira Takahashi and Masami Kawaguchi

Department of Industrial Chemistry, Faculty of Engineering, Mie University, Tsu, Mie 514, Japan

This article reviews recent advances in polymer adsorption both in theory and experiment. The adsorption of macromolecules on interfaces plays an essential role in the diversity of practical problems in industry, technology and biology such as adhesion, flocculation and stabilization of colloid particles, chromatography, reinforcement, and artifical organs in medicine.

This review appeals to researchers in the above mentioned fields and helps not only experimentalists but also theoreticians who are interested in polymer adsorption.

A. Introduction 3

B. Theories of Polymer Adsorption 4

B.1 Isolated Polymer Chains 5

B.2 Number of Configurations of a Tail and a Loop 6

B.3 Interacting Non-Ionic Polymer Chains 6

B.3.1 The Theory of Hoeve 6

B.3.1.1 Adsorption Isotherms 6

B.3.1.2 Segment Distribution Function for Loops 8

B.3.1.3 Excluded Volume Effect 10

B.3.2 The Theory of Silberberg 11

B.3.3 The Theory of Scheutjens and Fleer 16

B.3.4 Root-Mean-Square Thicknesses of Loops and Tails 25

B.3.4.1 Root-Mean-Square Thickness of Loops 25

B.3.4.2 Root-Mean-Square Thickness of Tails 26

B.3.5 Diffusion Equation Approach 26

B.3.6 The Scaling Theory 28

B.4 Theories of the Adsorption of Polyelectrolytes 30

B.4.1 The Theory of Frisch and Stillinger 30

B.4.2 The Theory of Hesselink 30

B.5 Summary 34

C. Experimental Techniques 35

C.1 Adsorption Isotherms 35

C.2 Thickness of the Adsorbed Layer 35

C.2.1 Hydrodynamic Methods 35

C.2.2 Ellipsometry 35

Advances in Polymers Science 46

C.2.3 The ATR Method 36
C.3 Fraction of Adsorbed Segments and Fraction of Occupied Surface Sites . 36
C.4 Heat of Adsorption 37

D. Experimental Results 37
D.1 Adsorption of Flexible Non-Ionic Polymers 37
D.1.1 Ellipsometric Studies 37
D.1.1.1 Adsorption at the Theta Point 37
D.1.1.2 Adsorption from Good Solvents 42
D.1.2 Hydrodynamic Studies 44
D.1.3 Other Approaches 46
D.1.4 Attachment of Train Segments to Active Sites 47
D.1.5 Heats of Adsorption 52
D.2 Adsorption of Branched Polymers 52
D.3 Adsorption of Block Copolymers 53
D.4 Adsorption of Polyelectrolytes 54
D.5 Adsorption of Block Polyelectrolytes 59
D.6 Adsorption of Polyampholytes 60
D.7 Adsorption of Rod-Like Macromolecules 60

E. Conclusions 61

F. References 62

A. Introduction

The adsorption of macromolecules on interfaces plays an essential role in the diversity of practical problems in industry, technology and biology, such as adhesion, flocculation and stabilization of colloid particles, chromatography, reinforcement, and artificial organs in medicine. Its features are quite different from those of the adsorption of small molecules. For example, the number of conformations displayed by a flexible polymer at an interface increases tremendously with chain length. Figure 1 depicts various chain conformations of adsorbed flexible and rod-like polymers. These conformations first determine the dimensions or thickness of the adsorbed polymer normal to the surface and second the configurational entropy and enthalpy of the adsorbed polymers. The free energy should be negative for the adsorption to occur. It is the fundamental property needed to calculate the extent of adsorption. Thus, the determination of conformations of adsorbed macromolecules is the central issue for polymer adsorption studies.

In 1953, Frisch, Simha, and Eirich[1–6] first investigated the change in conformation that occurs when a Gaussian coil is brought in contact with a reflecting wall, calculated thermodynamic properties of the adsorbed layer, and deduced an adsorption isotherm (FSE isotherm). *The important conclusion was that the thickness of the adsorbed layer at the theta point is proportional to the square root of the molecular weight of polymer.* However, in the early 1960's, their approach was criticized by Silberberg[7] and by DiMarzio[8], who showed that Frisch et al. had overestimated the number of distinguishable conformations of the adsorbed polymer chain. DiMarzio and McCrackin[9] showed that, for the correct evaluation of the number of conformations, an adsorbing barrier must be assumed just one step behind the wall.

In the mid-1960's extensive theoretical investigations of polymer adsorption began. The earlier studies were chiefly concerned with the conformation of isolated polymers,

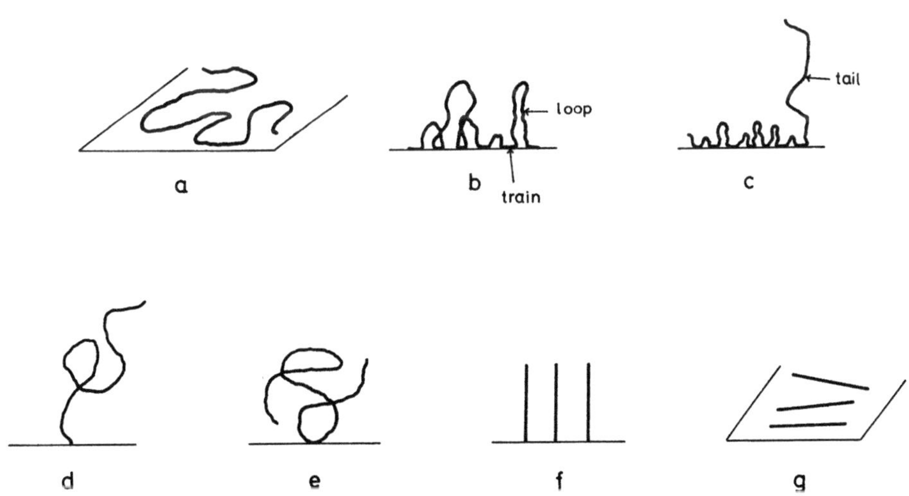

Fig. 1. Various conformation models for macromolecules adsorbed on an interface. **a**) chain lying totally on the interface; **b**) loop-train conformation; **c**) loop-train-tail conformation; **d**) adsorbed at one chain end; **e**) random coil adsorbed at a single point; **f**) rod-like macromolecules adsorbed at one end; **g**) rod-like macromolecules adsorbed located totally on an interface

the distributions of loops, trains, and tails, and the average thickness of adsorbed chain, statistical mechanical approaches[8, 10–14], lattice model approaches[7, 15, 16], and computer simulations by the Monte Carlo methods[9, 17–21] being used.

Although these studies were of great theoretical value, the results had little practical value since isolated adsorbed chains are inaccessible to experimental studies. The only exception was electron microscopic visualization of the conformation of a single DNA chain[22].

Subsequently, there emerged a number of extensions which aimed at incorporating lateral interactions of adsorbed polymers into the theory. The main interest was to predict how the adsorbance (mass per unit area) (A), the adsorbed amount (total number of segments adsorbed per site) (Γ), the fraction of segments in trains (p), and the thickness of the adsorbed layer depend on such physical variables as polymer molecular weight and thermodynamic interaction parameters between polymer-solvent and between polymer-adsorbing surface. Attemps have also been made to formulate the adsorption theory of polyelectrolytes[23] and to apply the scaling theory to polymer adsorption[24].

Concerning the experimental side of polymer adsorption studies the quantity A was only measurable at the early stage of the study, but in 1955 the thickness of the adsorbed layer became accessible to measurement by a hydrodynamic method and in 1961 the quantity p was first determined by infrared spectroscopy. Ellipsometry came up in 1963, which enabled both the adsorbance and the thickness of the adsorbed layer to be measured simultaneously.

No quantitative comparison between theory and experiment on polymer adsorption was attempted until the end of the 1970's. There were two reasons for this delay. First, no acceptable theory had been established. Second, some of the parameters used in most of the published theories could not be directly correlated to experimentally measurable quantities.

Polymer adsorption has been reviewed by many authors[28–37]. An earlier volume of this journal presented an article which dealt with polymer adsorption studies made before 1964[29]. This paper gives a review of subsequent advances in this field of study. In Chap. B, the principal theories are described, confining ourselves to those which are amenable to experimental tests. Chapter C gives a brief survey of typical measuring techniques. In Chap. D, important experimental data on the thickness of the adsorbed polymer layer and the fraction of adsorbed segments are summarized and discussed, along with their comparison with relevant theories.

B. Theories of Polymer Adsorption

B.1 Isolated Polymer Chains

It is well known that a free flexible polymer chain in bulk solution behaves as a random coil. When such a polymer chain is adsorbed on a surface, there occurs a change in its conformation. Some portions of the polymer chain get in direct contact with the surface as *trains*, and the remaining portions extend into the bulk solution as *loops* or *tails* (see Fig. 1).

Loops and tails of an isolated adsorbed polymer chain assume a number of different configurations and they substantially determine the configurational entropy of the adsorbed polymer, while the interaction energy between trains and the surface determines the enthalpy of adsorption.

Many authors[7–21] have theoretically investigated the conformation of an isolated adsorbed polymer as a function of adsorption energy, using statistical mechanical approaches. Some important conclusions are as follows:

1) At low adsorption energy, long loops or tails are favored and give rise to a conformation extended in the direction normal to the adsorbing surface, whereas at high adsorption energy, small loops or tails and long trains predominate, leading to a flattened conformation.
2) As intuitively expected, the number of train segments increases rapidly with increasing adsorption energy.
3) The distribution of loop segments is a simple exponential function of the distance from the surface while that of tail segments is the difference of two exponential functions and has a maximum at an intermediate distance.

B.2 Number of Configurations of a Tail and a Loop

The number of configurations $W_A(i, z)$ for a tail consisting of i segments that starts at the interface ($z = 0$) and ends at a distance z from the surface is given by[38–40]

$$W_A(i, z) = (2/\pi)^{1/2} 2^i z i^{-3/2} e^{-z^2/2i} \tag{B-1}$$

The total number of configurations $W_A(i)$ for this tail is obtained by integrating Eq. (B-1) over all allowable values of z, giving

$$W_A(i) = (2\pi i)^{-1/2} 2^i \tag{B-2}$$

A loop consisting of i segments can be formed by linking a tail of $(i - 1)$ segments that ends at $z = 1$ with the i-th segment on the interface. This i-th segment has only one possible orientation so that the number of configurations $W_{2A}(i)$ for a loop of i segments is equal to that of a tail of $(i - 1)$ segments that ends at $z = 1$. Thus, we have

$$W_{2A}(i) = W_A(i - 1, z = 1) \tag{B-3}$$

Substituting Eq. (B-1) into Eq. (B-3), we get

$$W_{2A}(i) = (2\pi)^{-1/2} 2^i i^{-3/2} \tag{B-4}$$

A comparison of Eqs. (B-2) and (B-4) shows that for the same i the tail formation is more favored than loop formation.

B.3 Interacting Non-Ionic Polymer Chains

Most of the early theories of polymer adsorption were not concerned with the interaction between adsorbed polymers so that they have little relevance for a comparison with experimental results. In actuality, the adsorbed mass per unit area is very large even when adsorption of polymers occurs from a very dilute solution. In this section, some typical theories allowing for the interaction between adsorbed polymers are reviewed.

B.3.1 The Theory of Hoeve

B.3.1.1 Adsorption Isotherms

Before describing this theory, we outline the theory of Hoeve et al.[41] formulated under the assumptions that the polymer chain is so long that end effects, i.e. tail formation, may be neglected and that the surface coverage is so low that the interaction of adsorbed polymer chains is negligible.

The following partition function q_a was derived for an isolated polymer chain consisting of trains and loops:

$$q_a = \Sigma(m!)^2 \prod_j (\sigma^{jm_j}/m_j!) \prod_i [(ci^{-3/2})^{n_i}/n_i!] \qquad \text{(B-5)}$$

where m is the number of trains, i.e. Σm_j, and equals the number of loops, i.e. Σn_i; σ is the internal partition function of a segment adsorbed on the surface relative to that in the bulk solution and c a flexibility parameter of the chain (for a flexible chain $c = 1$ and for a very stiff chain $c = 0$). The partition function for a loop of size i is obtained from Eq. (B-1). The total number of polymer segments, n, is expressed in terms of m_j and n_i as

$$n = \sum_j jm_j + \sum_i in_i \qquad \text{(B-6)}$$

By applying the Lagrangian multiplier method to Eq. (B-5), we obtain

$$n_i = mci^{-3/2} e^{i\lambda} e^{\xi} \qquad \text{(B-7)}$$

$$m_j = m\sigma^j e^{j\lambda} e^{-\xi} \qquad \text{(B-8)}$$

$$\ln q_a = -\lambda n \qquad \text{(B-9)}$$

where λ and ξ are multipliers and λkT may be regarded as the free energy of adsorption per segment. From Eq. (B-9) the following useful relationships can be derived.

1) The fraction of adsorbed segments, p:

$$p = \Sigma jm_j/n = (1 + S_{-3/2})/(1 + S_{-1/2} + S_{-3/2}) \qquad \text{(B-10)}$$

2) The average loop size, $\langle i \rangle$:

$$\langle i \rangle = \Sigma in_i/\Sigma n_i = S_{-1/2}/S_{-3/2} \qquad \text{(B-11)}$$

3) The average train size, $\langle m \rangle$:

$$\langle m \rangle = \Sigma j m_j / \Sigma m_j = S_{-3/2}^{-1} (1 + S_{-3/2}) \qquad \text{(B-12)}$$

where $S_{-\zeta} = \sum_i c i^{-\zeta} \exp(i\lambda)$. For a system of non-interacting N_a adsorbed and N_f unadsorbed polymers, the partition function Q is given by

$$Q = \Sigma[q_a{}^{N_a}(S\delta)^{N_a}/N_a!]\,[V^{N_f}/N_f!] \qquad \text{(B-13)}$$

where S is the area of the adsorbing surface, δ is the thickness of the first layer, i.e. the train layer, and V is the volume of the bulk solution. The total number of polymers, N, in the entire system is the sum of N_a and N_f:

$$N = N_a + N_f \qquad \text{(B-14)}$$

The adsorption isotherm is obtained by maximizing Eq. (B-13), which is subject to the constraint given by Eq. (B-14), yielding

$$N_a/S\delta = (N_f/V)\exp(-\lambda n) \qquad \text{(B-15)}$$

This expression predicts that the initial slope of a plot of ln $(N_a/S\delta)$ against ln (N_f/V) should be proportional to n. It is possible to calculate λ, p, m/n, the average loop size, and the average train size as functions of the two parameters σ and c. For $c = 1$ and 0.1, the average loop sizes are large and the average train sizes are small in the region $\sigma < 1$. In this region, p and m/n rapidly increase with σ. With a further increase of σ, p still increases and reaches unity at $\sigma = 3$, and a flattened conformation results. For a stiff chain, p is unity or zero, m/n is zero, and the train size becomes infinite. This means that all the segments of a stiff chain are either adsorbed completely or not at all. Therefore, the ability to form loops depends on the flexibility of the polymer and the loop size is determined by the free energy of adsorption.

The theory of Hoeve[42] was formulated by incorporating the interactions between adsorbed chains into the above theory of Hoeve et al.[41] The adsorbed layer is divided into two distinct regions: the first layer of thickness δ consists of segments in direct contact with the surface and the remaining layer consists of loops. The excess free energy ΔF_1 of interactions between the adsorbed segments in the first layer can be calculated from

$$\Delta F_1/kT = (S\delta/v_1)\,[(1 - \nu_2(0))\ln(1 - \nu_2(0)) + \nu_2(0)\,(1 - \chi\nu_2(0))] \qquad \text{(B-16)}$$

where k is the Boltzmann constant, T the absolute temperature, v_1 the volume occupied by a segment, $\nu_2(0)$ the volume fraction of segments in the first layer, i.e. $npv_1N_a/S\delta$, and χ the Flory-Huggins interaction parameter. The fraction p of adsorbed segments is given by

$$\Sigma j m_j = np \qquad \text{(B-17)}$$

With the restriction of Eq. (B-17) and using the Lagrangian multiplier method, we get for the free energy of adsorption ΔF_2

$$-\Delta F_2/kT = \ln q_a = n(-\lambda - p\eta) \tag{B-18}$$

where q_a is the partition function for the adsorbed polymer and both λ and η are multipliers. The total free energy of adsorption, ΔF, is the sum of ΔF_1 and ΔF_2, i.e.

$$\Delta F/kT = nN_a(\lambda + p\eta) + (S\delta/v_1)\,[(1 - \nu_2(0))\ln(1 - \nu_2(0)) + \nu_2(0)\,(1 - \chi\nu_2(0))] \tag{B-19}$$

Minimizing Eq. (B-19) with respect to p, we obtain

$$\eta \equiv [\ln(1 - \nu_2(0)) + 2\chi\nu_2(0)] = -\lambda - \ln\sigma - \ln(1 + S_{-3/2}) \tag{B-20}$$

The partition function for a system of N_a adsorbed and N_f free polymer molecules is given by

$$Q = N!\,\Sigma[\exp(-\Delta F/kT)(S\delta)^{N_a}/N_a!]\,(V^{N_f}/N_f!) \tag{B-21}$$

where $N = N_a + N_f$ and V is the volume of the solution. Applying the Lagrangian multiplier method, we obtain for the adsorption isotherm

$$N_a/S\delta = (N_f/V)\exp(-\lambda n) \tag{B-22}$$

which has the same form as Eq. (B-15) for non-interacting adsorbed polymers except for the parameter λ, the latter now depending on the adsorbance according to Eq. (B-20). Figure 2 shows isotherms for various values of n and the parameter values c = 0.1, $\sigma = 1.0$, and $\chi = 0.5$, i.e. the theta condition, as computed by Hoeve[42]. The adsorbance is here expressed in terms of a dimensionless quantity $nN_av_1/S\delta$. Even at very low bulk polymer concentration, is $nN_av_1/S\delta$ considerably large, increases gradually with rising polymer concentration, and levels off at relatively low concentration.

B.3.1.2 Segment Distribution Function for Loops

For a loop in which the first segment starts at the interface (z = 0) and the i-th segment first returns to the surface, Hoeve[43] derived the probability P(z, k; i) of finding its k-th segment at the distance z by use of a method analogous to that employed in deriving the end-to-end distribution of free chains. His formula is

$$P(z, k; i) = 4\pi^{-1/2}\beta_1^3 z^2\,[i/k(i - k)]^{3/2}\exp[-\beta_1^2 z^2 i/k(i - k)] \tag{B-23}$$

where $\beta_1^2 = 3n/2\,\langle r_n^2\rangle_0$, with $\langle r_n^2\rangle_0$ the mean-square end-to-end distance of an unperturbed chain with n segments. By combination with the loop size distribution, derived by Hoeve et al.[41], the normalized density distribution $\varrho_1(z)$ is obtained to be

$$\varrho_1(z) = n^{-1}\sum_i \sum_{k<i} n_i\,P(z, k; i) \tag{B-24}$$

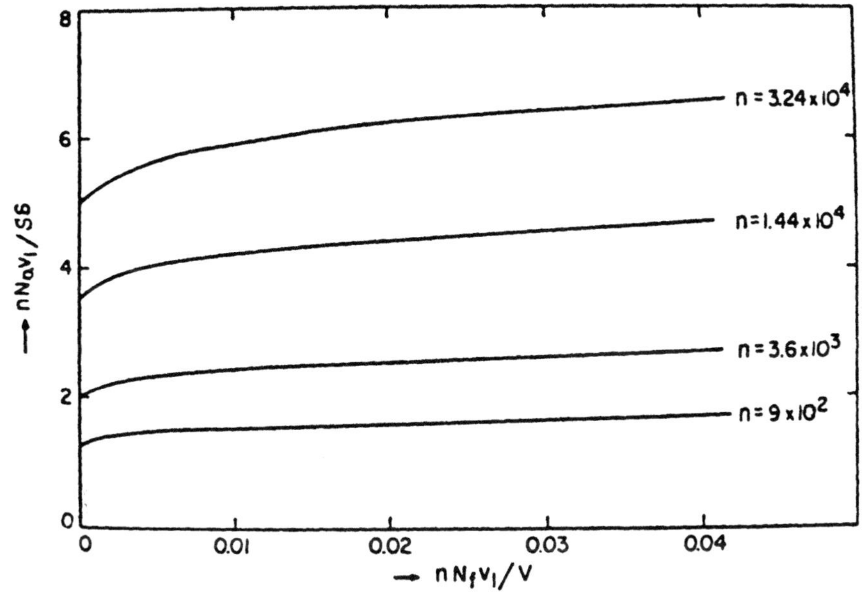

Fig. 2. Adsorption isotherms for different molecular weights[42]. The adsorbance $nN_av_1/S\delta$ is expressed in multiples of the amount that would fill the first layer. The polymer volume fraction in solution is given by nN_fv_1/V where n is the number of segments

where n_i is the number of loops of size i. Putting $n_i = mc_i^{-3/2} e^{i\lambda} e^{\xi}$, which is given by Eq. (B-7), and integrating Eq. (B-24) over z, we find

$$\varrho_1(z) = 4\pi^{1/2}\beta_1 c \exp[-4\beta_1 z(-\lambda)^{1/2}][1 + S_{-1/2} + S_{-3/2}]^{-1} \tag{B-25}$$

This result is valid for z larger than δ, the mean thickness of train segments that are in contact with the surface. Eq. (B-25) indicates that $\varrho_1(z)$ is a simple exponential function of z.

For a system with N_a adsorbed polymers, each having n segments, the volume fraction $\varrho_2(z)$ of segments at $z > \delta$ is given by

$$\varrho_2(z) = \varrho_1(z)\,(nN_av_1/S) \tag{B-26}$$

For small $|\lambda|$ Eq. (B-26) is rewritten, using Eq. (B-25), as

$$\varrho_2(z) = K\nu_2(0) \exp[-pKz/\delta\,(1 - p)] \tag{B-27}$$

where $K = 4\pi^{1/2}\beta_1 c(1 + 2.612c)^{-1}$. For $p \ll 1$, Eq. (B-27) reduces to

$$\varrho_2(z) = K\nu_2(0) \exp - (pKz/\delta) \tag{B-28}$$

The root-mean-square distance $\langle z^2\rangle^{1/2}$ of segments from the interface is calculated from

$$\langle z^2 \rangle = \int_0^\infty z^2 \varrho_1(z)\, dz \tag{B-29}$$

which gives, by substitution of Eq. (B-25),

$$\langle z^2 \rangle^{1/2} = 2^{-3/2} \beta_1^{-1} (-\lambda)^{-1/2} \tag{B-30}$$

Since $-\lambda n$ changes little with n for small $|\lambda|$, *the root-mean-square thickness of segments* $\langle z^2 \rangle^{1/2}$ *is approximately proportional to* $n^{1/2}$. In Fig. 3, the dimensionless quantity $\beta_1 \langle z^2 \rangle^{1/2}$ is plotted against $n^{1/2}$ for $c = 0.1$, $\sigma = 1.0$ and $\nu_2(0) = 0.04$ and 0.004[42]. The calculated curves are almost linear.

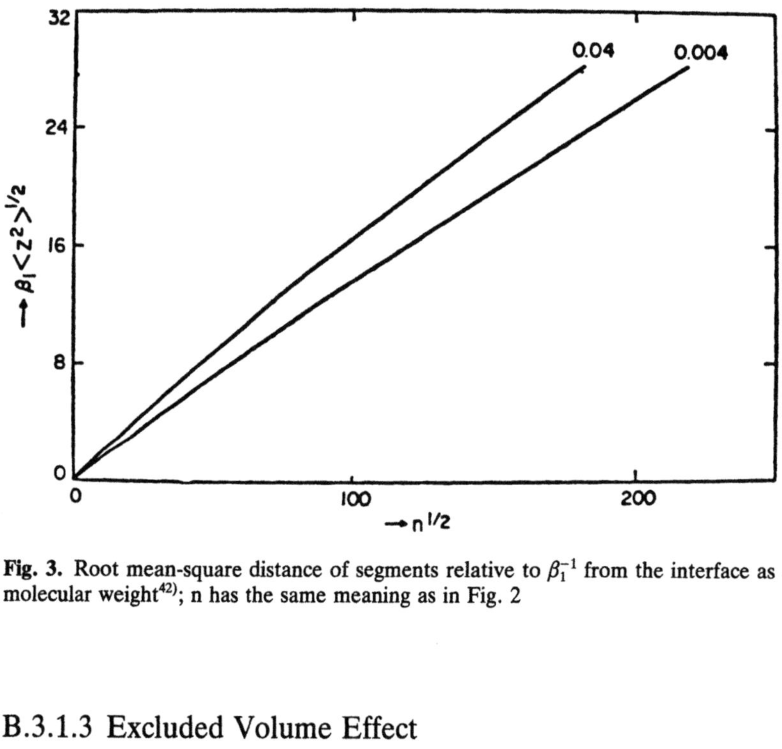

Fig. 3. Root mean-square distance of segments relative to β_1^{-1} from the interface as a function of molecular weight[42]; n has the same meaning as in Fig. 2

B.3.1.3 Excluded Volume Effect

Hoeve[44, 45] extended his theory further by considering not only interactions between the train segments but also interactions among the loops, and found that the latter lead to a decrease in the number of possible conformations of adsorbed polymer chains. He assumed that the segment density distribution in any loop is uniformly expanded in one dimension by a factor of α_t as a result of loop-loop interactions. The volume fraction of segments at a distance $z > \delta$ is then given by

$$\varrho_3(z) = K \alpha_t^{-1} \nu_2(0) \exp(-pKz/\delta\, \alpha_t) \tag{B-31}$$

The elastic free energy ΔF_{el} due to the expansion of mN_a loops is expressed by[46)]

$$\Delta F_{el}/kT = mN_a[3/2\,(\alpha_t^2 - 1) - 3\ln \alpha_t] \tag{B-32}$$

The free energy ΔF_{int} of interactions between segments in the loop layer can be calculated by using $\varrho_3(z)$ and the Flory-Huggins theory, yielding

$$\Delta F_{int} = \frac{1}{2}\, nN_a kT\left(\frac{1}{2} - \chi\right) K\alpha_t^{-1} \nu_2(0) \tag{B-33}$$

The total interaction free energy of loops ΔF_3 is the sum of ΔF_{el} and ΔF_{int}. Minimizing ΔF_3 with respect to α_t at constant values of N_a and p, we obtain for α_t

$$(\alpha_t^3 - \alpha_t) = \left(\frac{1}{6}\right) (m/n)^{-1}\left(\frac{1}{2} - \chi\right) K\nu_2(0) \tag{B-34}$$

Moreover, it can be shown that the adsorption isotherm for the system containing N_f polymer chains in the bulk solution of volume V and N_a chains in the adsorbed layer is given by

$$N_a/S\delta = (N_f/V)\,\exp\left\{-n\left[\lambda + \frac{1}{2}\left(\frac{1}{2} - \chi\right) K\nu_2(0)\right]\right\} \tag{B-35}$$

This isotherm reduces to Eq. (B-22) when $\chi = 1/2$ (the theta condition). According to Eq. (B-35), the adsorbance is proportional to the solution concentration, as is the case with non-interacting chains.

In the limit of infinitely high molecular weight we have

$$\lim_{n\to\infty}\,[(N_a/S\delta)/(N_f/V)]^{1/n} = 1 \tag{B-36}$$

Hence, Eq. (B-35) gives at this limit

$$\lambda + \left(\frac{1}{2} - \chi\right) K\nu_2(0)/2 = 0 \tag{B-37}$$

which predicts a maximum to occur in the adsorbance. As the solvent gets better, the limiting values of $(-\lambda)$ and p become larger, the average loop size defined by Eq. (B-30) becomes smaller, and the limiting adsorbance is given by Eq. (B-36).

B.3.2 The Theory of Silberberg

Silberberg[47)] used a quasi-crystalline lattice model for the adsorption of flexible macromolecules. If it is assumed that an adsorbed polymer chain with P segments consists of m_{si} trains of length i and m_{Bi} loops of length i, the total number of configurations of the chains is given by

$$\Omega_p^\sigma = \left[\prod_i \omega_{si}{}^{m_{si}}/m_{si}!\right]\left[\prod_i \omega_{Bi}{}^{m_{Bi}}/m_{Bi}!\right]\left(\sum_i m_{si}!\right)\left(\sum_i m_{Bi}!\right) \tag{B-38}$$

where

$$\sum_i i\,m_{si} + \sum_i i\,m_{Bi} = P \tag{B-39}$$

$$\sum_i m_{si} = \sum_i m_{Bi} \tag{B-40}$$

and ω_{si} and ω_{Bi} are the numbers of available configurations of trains and loops, respectively, and defined by

$$\omega_{si} = \gamma_s\, i^{-a_s}(U_s)^i \tag{B-41}$$

$$\omega_{Bi} = \gamma_B\, i^{-a_B}(U_B)^i \tag{B-42}$$

In these expressions, U_B is the number of ways of adding one segment to an essentially infinite chain or one loop in the three-dimensional bulk phase, U_s is the corresponding number for a two-dimensional train on the surface, a_B and a_s are excluded-volume parameters in the bulk and on the surface, respectively, and γ_B and γ_s the probability of transforming a sequence of adsorbed segments to a loop and that of the reverse process, respectively. The partition function $q^\sigma(P, T)$ of the system is obtained from Eq. (B-38) as follows:

$$q^\sigma(P, T) = q^*(P, T)(1/\Omega_p^*) \sum_{\{m_{si}, m_{Bi}\}} \Omega_p^\sigma(m_{si}, m_{Bi}) \exp\left(\sum_i i\,m_{si}\chi_s\right) \tag{B-43}$$

where q^* is the partition function of an isolated polymer in the bulk solution, χ_s the energy in kT units required to replace a solvent molecule on the surface by a polymer segment, and Ω_p^* the number of arrangements possible for the polymer in the bulk solution.

From the maximum term of Eq. (B-43), taking into account, Eqs. (B-39) and (B-40) the average loop length P_B, the average train length P_s, and the fraction p of adsorbed segments can be calculated as functions of γ_s, γ_B, a_s, a_B, and ΔF_s. The last quantity is related to χ_s by

$$-(\Delta F_s/kT) = \ln(U_s/U_B) + \chi_s \tag{B-44}$$

The surface layer is assumed to be a two-dimensional lattice of coordination number C_s, containing M_s sites occupied by either solvent molecules or polymer segments. When N_a polymers are adsorbed and M_{sP} segments are attached to the surface, the surface coverage θ is given by

$$\theta = M_{sP}/M_s \tag{B-45}$$

and M_{sP} can be written in terms of the number of trains m_{si} as

$$M_{sP} = \left(\sum_i i\, m_{si}\right) N_a \tag{B-46}$$

The number of segments M_{BP} in the loop is expressed by

$$M_{BP} = \left(\sum_i i\, m_{Bi}\right) N_a = PN_a - M_{sP} \tag{B-47}$$

The thickness of the adsorbed region B which consists of a number of layers, each with M_s sites, is assumed to be $\delta' P_B^{\beta'}$ where δ' and β' are constants. Silberberg[47)] assumes that $\delta' = 1/2$ and $\beta' = 1$. Beyond this region is there a homogeneous bulk solution of polymer volume fraction ϕ^* from which adsorption has taken place. Except for solvent, all that is found in the region B are loop segments of the adsorbed polymer chains. Thus, the fraction ϕ_B of adsorbed polymer segments is expressed by

$$\phi_B = M_{BP}/M_B \tag{B-48}$$

which can be rewritten

$$\phi_B = (P_B/P_s)\, \theta\, (1/\delta' P_B^{\beta'}) \tag{B-49}$$

This equation represents a stoichiometric relation between the volume fraction ϕ_B in the region B and the surface coverage θ.

The same coordination number C_B is assumed for both the region B and the bulk solution. The change C_{sB} in the contact number which occurs when the adsorbed polymer chain goes from M_s surface sites to the region B is expressed by

$$C_{sB} = \frac{1}{2}(C_B - C_s) \tag{B-50}$$

The corresponding change C'_{sB} in going from M_B sites in the region B to the surface is expressed by

$$C'_{sB} = C_{sB}(M_s/M_D) \tag{B-51}$$

Therefore, the effective coordination number C'_B in the region B is

$$C'_B = C_B - C'_{sB} = C_B - \frac{C_{sB}}{2}(C_B - C_s) \tag{B-52}$$

In addition to this expression for C'_B we need to know the configurational factor and the configurational energy in order to calculate the partition function. The configurational energy is the sum of the interaction energies associated with segment-segment contacts (p, p), segment-solvent contacts (p, o), and solvent-solvent contacts (o, o). The interaction parameter χ is related to the coordination number C_B by

$$\chi = (C_B/kT)\left[V_{op} - \frac{1}{2}(V_{pp} + V_{oo})\right] \tag{B-53}$$

where V_{op}, V_{pp}, and V_{oo} are the potential energies for the contacts (o, p), (p, p), and (o, o), respectively. The potential energy for the contact between the solvent and the surface and that for the contact between the segment and the surface are denoted by V_{so} and V_{sp}, respectively. In terms of these parameters the segment-surface interaction parameter χ_s is expressed by

$$\chi_s = (C_{sB}/kT)\,[V_{so} - (V_{sp} - V_{pp}) - \frac{1}{2}(V_{oo} + V_{pp})] \qquad \text{(B-54)}$$

The partition function can then be written

$$Q^{total} = \sum_{\{m_{si}, m_{Bi}, N_o^*, N_a^*, N_a\}} [g^{\sigma}/g_p^R(N_{RP})g_o^R(N_{Ro})]\,\eta \qquad \text{(B-55)}$$

where N_o^* and N_a^* are the numbers of solvent and polymer molecules in the bulk solution, g^{σ} is the configurational factor for the polymer solution, $g_p^B(N_{RP})$ and $g_o^R(N_{Ro})$ are the configurational factors for the reference state consisting of N_{RP} pure polymer molecules and N_{Ro} solvent molecules, and η is related to the energy of mixing ΔU by

$$\Delta U = -kT\ln\eta \qquad \text{(B-56)}$$

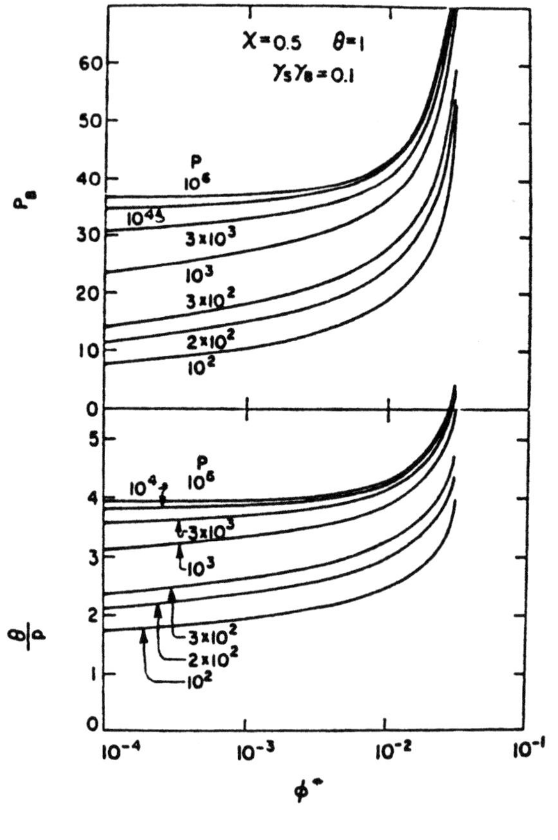

Fig. 4. Average loop length P_B (proportional to film thickness) and adsorbed amount θ/p as a function of equilibrium polymer concentration ϕ^* and degree of polymerization P for $\gamma_s\gamma_B = 0.1$, $\chi = 0.5$, and $\chi_s = \infty\ (\theta = 1)$[47)]

The above theory does not allow analytical expressions to be obtained for the adsorption isotherm and the thickness of the adsorbed layer. Using the maximum term in Eq. (B-55), Silberberg[47] numerically evaluated N_o^*, N_a^*, N_a, m_{si}, and m_{Bi}, though the computation was limited to two special cases of χ: $\chi = 0$ (athermal) and $\chi = 1/2$ (the theta point). With the lattice coordination number and the number of possible arrangements of polymer held fixed, Silberberg[47] computed the values of P_B, P_s, θ, p, and ϕ_B by varying the parameters ϕ^*, P, $\gamma_s\gamma_B$, and χ_s. Some of his results are shown in Figs. 4–7. At fixed P and $\gamma_s\gamma_B$ (the latter represents the flexibility of polymer), the χ_s dependence of the adsorbed amount, i.e. the average number of adsorbed segments per surface site, θ/p, showed features that θ/p for $\chi = 0.5$ was twice as high as that for $\chi = 0$ and practically independent of χ_s above some critical value of χ_s which corresponds to the onset of polymer adsorption. The corresponding χ_s dependence of the average loop size P_B, which is proportional to the thickness of the adsorbed layer, was similar to that of θ/p. At the theta point, the dependence of P_B and θ/p on ϕ^* and P was most appreciable, but, under athermal conditions, P_B and θ/p were almost independent of ϕ^* and P (see Figs. 4 and 5). From Fig. 6 showing the calculated results for $\chi = 1/2$, we see that *P_B at the theta point is approximately proportional to the square root of molecular weight over a wide range, but approaches a constant for large molecular weights.* On the other hand, the P dependence of P_B under athermal conditions is weak, as shown in Fig. 7.

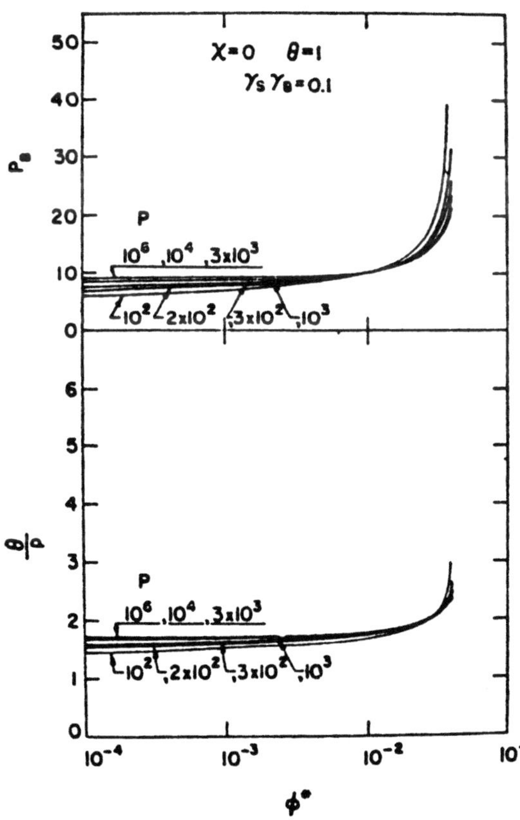

Fig. 5. The same as in Fig. 4 except for $\chi = 0$[47]

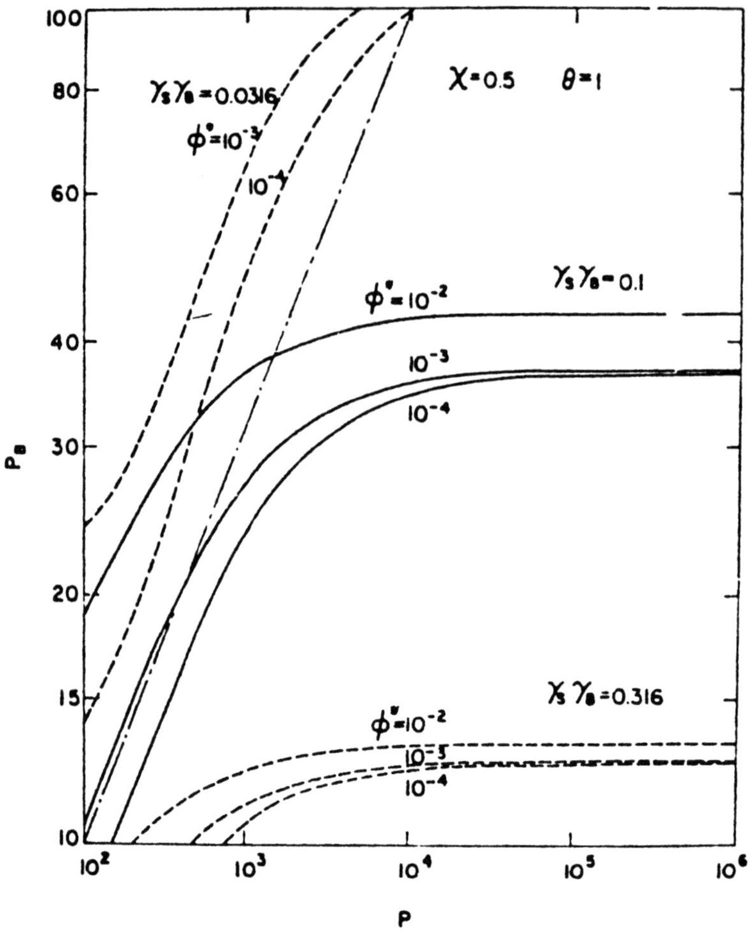

Fig. 6. Dependence of average loop length P_B (proportional to film thickness) on degree of polymerization P for $\chi = 0.5$, $\chi_s = \infty$ ($\theta = 1$) for $\gamma_s\gamma_B = 0.0316$, 0.1, 0.316. and $\phi^* = 10^{-4}$, 10^{-3}, 10^{-2} [47]. The straight line represents the relation $P_B \propto P^{1/2}$

B.3.3 The Theory of Scheutjens and Fleer

In the theory of Scheutjens and Fleer[48], the system is a mixture of N polymer chains, each consisting of r segments, and N° solvent molecules, and the region adjacent to the adsorbing surface is divided into M layers parallel to the surface, each containing L lattice sites. Thus, we have the relation

$$N^\circ + rN = ML \tag{B-57}$$

If the volume fractions of solvent and polymer in the i-th layer are denoted by ϕ_i° and ϕ_i, respectively, we have

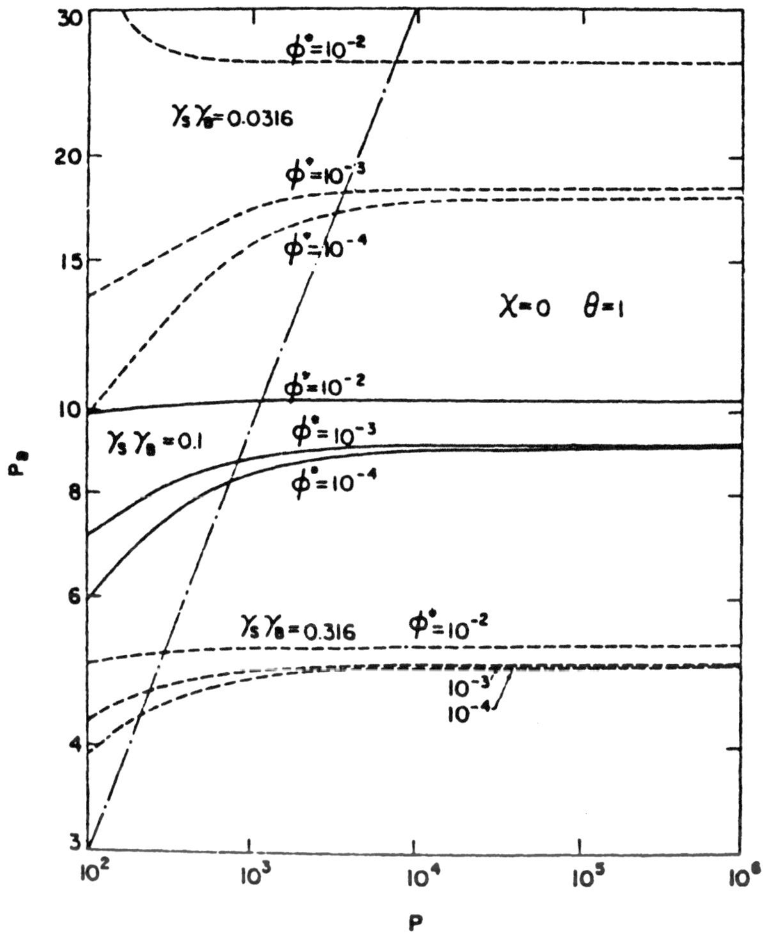

Fig. 7. The same as in Fig. 6 except for $\chi = 0$[47)]

$$\phi_i^\circ = N_i^\circ/L, \qquad \phi_i = N_i/L \tag{B-58}$$

with N_i° and N_i the numbers of solvent molecules and polymer segments in the i-th layer, respectively. Each lattice site has z nearest neighbors, $z\lambda_0$ in the same layer and $z\lambda_1$ in either of the adjacent layers. Therefore, we have

$$\lambda_0 + 2\lambda_1 = 1 \tag{B-59}$$

Each adsorbed chain assumes a large number of conformations. A conformation can be defined by the layer number in which a specified segment finds itself. Thus, it is symbolically represented by

$$(1, i)\,(2, j)\,(3, k)\cdots(r-1, 1)\,(r, m) \tag{B-60}$$

when the first segment is in the i-th layer, the second in the j-th layer, the third in the k-th layer, and so on. If this conformation is designated as conformation c, the number of different arrangements of segment in it is given by

$$Lz^{r-1}\lambda_{j-i}\lambda_{k-j}\cdots\lambda_{m-1} = Lz^{r-1}\omega_c = Lz^{r-1}\prod_{s=1}^{r-1}(\lambda_{s,s+1})_c \tag{B-61}$$

where $(\lambda_{s,s+1})_c = \lambda_0$ if segments s and s + 1 are in the same layer and $(\lambda_{s,s+1})_c = \lambda_1$ if these two segments are in the neighboring layers. Since z^{r-1} is the number of arranging segments for a chain in bulk solution, ω_c may be considered the number of arranging segments in conformation c relative to that in bulk solution.

If the number of polymer molecules which assumes conformation c is N_c, N is given by

$$N = \sum_c N_c \tag{B-62}$$

Denoting by $r_{i,c}$ the number of segments in the i-th layer associated with conformation c, we can express the total number of segments N_i in the i-th layer by

$$N_i = \sum_c r_{i,c}N_c \tag{B-63}$$

This corresponds to one particular way of grouping the possible arrangements of individual chains into one set.

The partition function Q(M, L, T, $\{N_c\}$) for a specified set of conformations $\{N_c\}$ is given by

$$Q(M, L, T, \{N_c\}) = (\Omega/\Omega^+)\,e^{-G/kT} \tag{B-64}$$

where Ω is the number of arrangements of polymer chains in the specified conformation, Ω^+ the number of arrangements of polymer chains over Nr lattice sites in the amorphous bulk polymer, and G the energy of the system. The combinatory factor Ω is expressed by

$$\Omega = [(L!)^M(z/L)^{(r-1)N}]\prod_c \omega_c^{N_c}/N_c!\prod_{i=i}^{M} 1/N_i^\circ! \tag{B-65}$$

The combinatory factor Ω^+ can be written

$$\Omega^+ = [(rN)!/N!](z/rN)^{(r-1)N} \tag{B-66}$$

If Eqs. (B-65) and (B-66) are substituted into Eq. (B-64), the entropy part of Q is found, by applying the Stirling's approximation, to be

$$\ln(\Omega/\Omega^+) = ML\ln L - \sum_c N_c\ln(N_c/\omega_c) - \sum_i N_i^\circ\ln N_i^\circ - N\ln r - (r-1)N\ln L \tag{B-67}$$

If the Flory-Huggins relation is used, the energy part of Q is given by

$$U = N_1 u_s + N_1^\circ u_s^\circ + kT\chi \sum_{i=1}^{M} N_i^\circ \langle \phi_i \rangle \tag{B-68}$$

where u_s and u_s° are the adsorption energies of a segment and a solvent molecule, respectively, χ is the Flory-Huggins parameter, and the site volume fraction $\langle \phi_i \rangle$ of polymer segments is defined by

$$\langle \phi_i \rangle = \sum_{j=1}^{M} \lambda_{j-i} \phi_i \tag{B-69}$$

In order to find the number of chains N_d in a particular conformation d, ln Q is differentiated with respect to N_d, and we get

$$\ln(N_d/L) = \ln a + \ln \omega_d + \sum_{i=1}^{M} r_{i,d} \ln P_i \tag{B-70}$$

or

$$N_d/L = a\,\omega_d \prod_{i=1}^{M} P_i^{r_{i,d}} \tag{B-71}$$

where a is a constant, $r_{i,d}$ the number of solvent molecules displaced when one polymer chain in conformation d is added to the i-th layer, and P_i the probability of finding a free segment in the i-th layer and defined by

$$\ln P_i = \chi_s \delta_{1,i} + (\langle \phi_i \rangle - \langle \phi_i^\circ \rangle) + \ln \phi_i^\circ \tag{B-72}$$

with the adsorption energy parameter χ_s given by $-(u_s - u_s^\circ)/kT$, $\delta_{1,i}$, the Kronecker delta, and $\langle \phi_i^\circ \rangle$ the site volume fraction of the solvent in the i-th layer.

The conformation probability $P(r)_c$ is represented by

$$P(r)_c = \omega_c \prod_{i=1}^{M} P_i^{r_{i,c}} = \omega_c \prod_{s=1}^{r} P_{k(s,c)} \tag{B-73}$$

where $P_{k(s,c)}$ is the probability that segment s of a chain in conformation c is found in the k-th layer. The chain probability P(r) is the sum of $P(r)_c$ over all possible conformations:

$$P(r) = \sum_{c} P(r)_c \tag{B-74}$$

It can be also expressed by

$$P(r) = \sum_{i=1}^{M} P(i,r) \tag{B-75}$$

if P(i,r) is the probability that one of the end segments of a chain with r segments is found in the i-th layer. Thus, we obtain the relation

$$\phi_i = (N/LP(r)P_i) \sum_{s=1}^{r} P(i,s)\,P(i,r-s+1) \tag{B-76}$$

which may be rewritten in terms of the polymer volume fraction $\phi_* = Nr/LP(r)$ in bulk solution as

$$\phi_i = (\phi_*/rP_i) \sum_{s=1}^{r} P(i,s)\, P(i,r-s+1) \tag{B-77}$$

The quantity ϕ_i can then be calculated by the matrix procedure, and the adsorbed amount Γ is obtained from

$$\Gamma = \sum_{i=1}^{M} (\phi_i - \phi_i^f) \tag{B-78}$$

where ϕ_i^f is the volume fraction of unadsorbed chains in the i-th layer.

The adsorbed amount Γ, the surface coverage θ, and the fraction of adsorbed segments $p = \theta/\Gamma$ have been computated by Scheutjens and Fleer[48] as functions of ϕ_*, r, and χ_s for two χ parameters; $\chi = 1/2$ and $\chi = 0$. Most of the computations have been made for the closest packed hexagonal lattice. The resulting values of Γ, θ, and p plotted against ϕ_* are illustrated in Figs. 8a, 8b, and 8c, respectively. The adsorption isotherms are similar to the Langmuir type for low molecular weights, but they are of the high

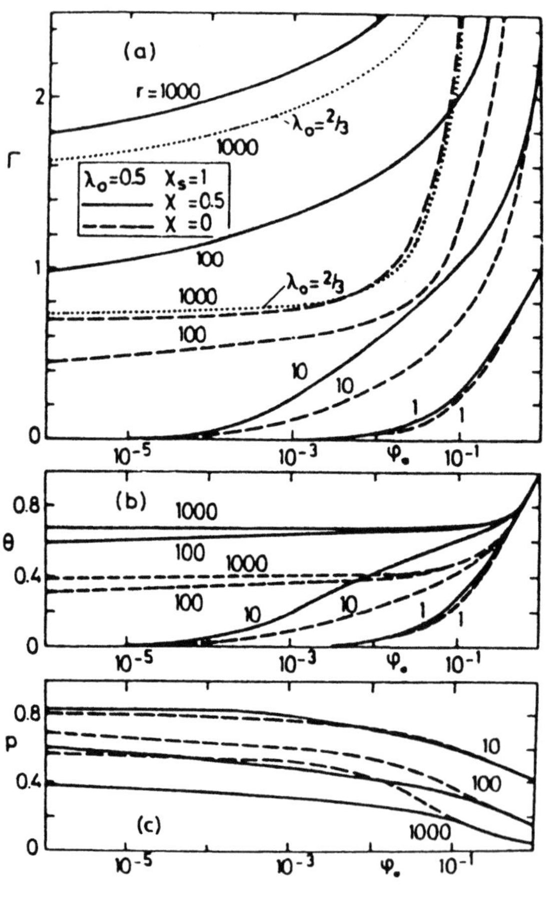

Fig. 8. (**a**) Adsorbed amount Γ, (**b**) surface coverage θ, and (**c**) fraction of adsorbed segments $p = \theta/\Gamma$ as functions of polymer volume fraction ϕ_*[48]. The data refer to the hexagonal lattice ($\lambda_0 = 0.5$) and the adsorption energy parameter $\chi_s = 1$. Chain length r is indicated. Full lines are for a θ-solvent ($\chi = 0.5$), broken lines for an athermal solvent ($\chi = 0$). Dotted lines in (a) relate to r = 1,000 on the simple cubic lattice ($\lambda_0 = 2/3$) and $\chi = 0$ and 0.5

affinity type for high molecular weights. For $\chi = 0$, Γ increases with increasing chain length and levels off for very large chains. This molecular weight dependence of Γ is in agreement with the predictions of Silberberg[47] and Hoeve[42]. On the other hand, for $\chi = 0.5$, Γ increases linearly with log r. Thus, the results for theta solvents support Hoeve's predictions[42] though Hoeve's theory ignores the contribution of tail portions. Γ increases with rising χ_s and becomes constant around $\chi_s = 3$, both for $\chi = 0.5$ and 0. This χ_s dependence of Γ very much resembles that predicted by Roe's treatment[49] based on a lattice model, but we note that Roe's theory assumes no particular conformation for adsorbed chains.

The surface coverage θ increases with rising ϕ_* for short chains, but it is nearly independent of ϕ_* for long chains. The values of θ under theta conditions are larger than those under athermal conditions. As the chain length grows, these values increase and reach a constant independent of ϕ_*. They also increase with rising χ_s in the region below 3.0. This tendency is in agreement with the predictions of Roe[49].

The fraction of adsorbed segments, p, decreases slightly with increasing ϕ_* and appreciably with growing molecular weight. Moreover, it diminishes with decreasing χ for long chains.

Scheutjens and Fleer further calculated[50] the concentration distributions of loop and tail segments, the root-mean-square thickness of the adsorbed layer, and the average numbers and lengths of trains, loops, and tails. They also computed the distributions of trains, loops, and tails from the free segment probability P_i, which may be represented by

$$P_i = (\phi_i^\circ/\phi_*^\circ) \exp[2\chi(\langle\phi_i\rangle - \phi_*)] \exp(\chi_s + \lambda_1\chi)\,\delta_{1,i} \qquad \text{(B-79)}$$

Here the factor $\phi_i^\circ/\phi_*^\circ$ is associated with the difference in the configurational entropy of a segment between the i-th layer and the bulk solution and the term $\exp[2\chi(\langle\phi_i\rangle - \phi_*)]$ with the interaction between a segment and its nearest neighbors. We note that with Eq. (B-69), the site volume fraction $\langle\phi_i\rangle$ is written

$$\langle\phi_i\rangle = \lambda_1\phi_{i-1} + \lambda_0\phi_i + \lambda_1\phi_{i+1} \qquad \text{(B-80)}$$

The term $\exp(\chi_s + \lambda_1\chi)\delta_{1,i}$ corresponds to the adsorption free energy.

The overall segment density depends on the number of chains in a particular conformation. Hence, the volume fraction $\phi_i(s)$ of segment $s(s = 1, 2, \ldots, r)$ is expressed in terms of the probability P(s, i; r) that the s-th segment of an r-mer is found in the i-th layer as follows:

$$\phi_i(s) = (\phi_*/r)\, P(s, i; r) \qquad \text{(B-81)}$$

Since P(s, i; r) is given by

$$P(s, i; r) = P(i, s)\, P(i, r - s + 1)/P_i \qquad \text{(B-82)}$$

Eq. (B-81) is rewritten

$$\phi_i(s) = (\phi_*/rP_i)\, P(i, s)\, P(i, r - s + 1) \qquad \text{(B-83)}$$

Thus, the overall volume fraction ϕ_i is expressed by

$$\phi_i = (\phi_* / rP_i) \sum_{s=1}^{r} P(i, s) P(i, r - s + 1) \quad \text{(B-84)}$$

In general, P(i, s) can be represented as

$$P(i, s) = P_a(i, s) + P_f(i, s) \quad \text{(B-85)}$$

with $P_a(i, s)$ and $P_f(i, s)$ representing the end segment probabilities for adsorbed and free chains, respectively. Equation (B-84) can be solved by the matrix procedure to obtain ϕ_i.

The quantity ϕ_i is the sum of the contributions from free chains ϕ_i^f and adsorbed chains ϕ_i^a. The latter is the sum of the contributions of trains (ϕ_i^{tr}), loops (ϕ_i^{ℓ}), and tails (ϕ_i^t). Since the segments in the first layer belong to trains, ϕ_1 is written

$$\phi_1 = \phi_1^{tr} \quad \text{(B-86)}$$

For i > 1, ϕ_i is represented by

$$\phi_i = \phi_i^{\ell} + \phi_i^t + \phi_i^f \quad \text{(B-87)}$$

with ϕ_i^{ℓ}, ϕ_i^t, and ϕ_i^f defined by

$$\phi_i^{\ell} = (\phi_* / rP_i) \sum_{s=1}^{r} P_a(i, s) P_a(i, r - s + 1) \quad \text{(B-88)}$$

$$\phi_i^t = (2\phi_* / rP_i) \sum_{s=1}^{r} P_a(i, s) P_f(i, r - s + 1) \quad \text{(B-89)}$$

$$\phi_i^f = (\phi_* / rP_i) \sum_{s=1}^{r} P_f(i, s) P_f(i, r - s + 1) \quad \text{(B-90)}$$

From these we can evaluate the concentration distribution of segments in the adsorbed layer.

The overall root-mean-square thickness t_{rms} can be represented in terms of ϕ_i^a as

$$t_{rms}^2 = (1/\Gamma) \sum_{i=1}^{M} i^2 \phi_i^a \quad \text{(B-91)}$$

where the adsorbed amount Γ is given by $\sum_{\iota=1}^{M} \phi_i^a$. Analogously, the root-mean-square thicknesses t_{ℓ} and t_t for loops and tails are defined by

$$t_{\ell}^2 = \sum_{i=1}^{M} i^2 \phi_i^{\ell} / \sum_{i=1}^{M} \phi_i^{\ell} \quad \text{(B-92)}$$

$$t_t^2 = \sum_{i=1}^{M} i^2 \phi_i^t / \sum_{i=1}^{M} \phi_i^t \quad \text{(B-93)}$$

The number of loops per chain n_ℓ is expressed by

$$n_\ell = (\lambda_1/P_a(r)) \sum_{s=1}^{r-1} P_a(2, s)\, P_a(1, r-s) \tag{B-94}$$

The number of trains per chain n_{tr} is calculated from

$$n_{tr} = n_\ell + 1 \tag{B-95}$$

In the same way, the number of tails per chain n_t is

$$n_t = \sum_{s=1}^{r-1} n_t(s) = (2\lambda_1/P_a(r)) \sum_{s=1}^{r-1} P_f(2, s)\, P_a(1, r-s) = 2 - 2\,P_a(1, r)/P_a(r) \tag{B-96}$$

The fractions of segments in trains (ν_{tr}), tails (ν_t), and loops (ν_ℓ) are defined by

$$\nu_{tr} = \phi_1/\Gamma \tag{B-97}$$

$$\nu_t = (1/r) \sum_{s=1}^{r-1} s\, n_t(s) \tag{B-98}$$

and

$$\nu_\ell = 1 - \nu_{tr} - \nu_t \tag{B-99}$$

Moreover, the average lengths of trains, tails, and loops can be calculated from

$$\ell_{tr} = r\nu_{tr}/n_{tr} \tag{B-100}$$

$$\ell_t = r\nu_t/n_t \tag{B-101}$$

and

$$\ell_\ell = r\nu_\ell/n_\ell \tag{B-102}$$

Figure 9 shows the volume fractions of tail, loop, and overall segments as functions of the distance i from the surface for a polymer chain consisting of 38% train segments, 55.5% loop segments, and 6.5% tail segments, as computed by Scheutjens and Fleer[50]. It is clear that the concentration of loop segments decreases exponentially with i. This result is in good agreement with the theoretical predictions of Hoeve[42], Roe[12], Silberberg[47], and Hesselink[39, 40]. Since the tail segments dominate at distances far from the surface, the overall segment distribution is not a simple exponential.

Scheutjens and Fleer[50] computed the root-mean-square thickness t_{rms} as a function of $\sqrt{r}$. The result is shown in Fig. 10. *For* $\chi = 1/2$ t_{rms} *is proportional to* $\sqrt{r}$ above r = 20, while for $\chi = 0$, this quantity is not completely linear against $\sqrt{r}$ and its magnitude is somewhat smaller than that for $\chi = 0.5$. *Moreover, though not illustrated here, both the root-mean-square thickness of tail segments,* t_t*, and that of loop segments,* t_ℓ*, are approximately proportional to* $\sqrt{r}$*,* but the former is much larger than the latter. The linear relation between t_ℓ and $\sqrt{r}$ agrees well with the prediction by Hoeve's theory[42].

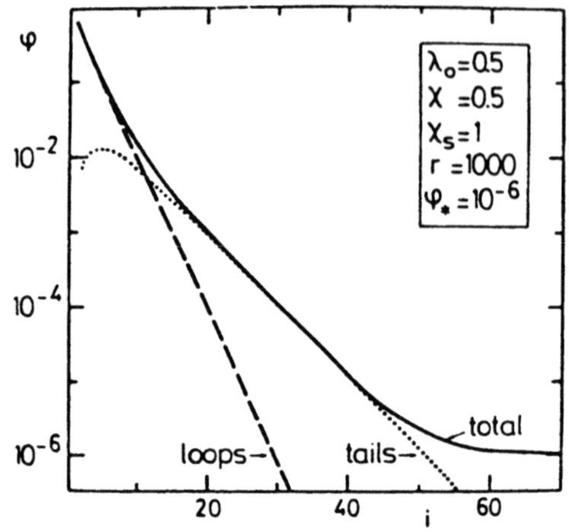

Fig. 9. Volume fraction ϕ_i (full line) and its components ϕ_i^ℓ for loops (broken line) and ϕ_i^t for tails (dotted line) on the logarithmic scale as functions of the distance from the surface[50]. The loop contribution ϕ_i^ℓ decreases essentially exponentially with i

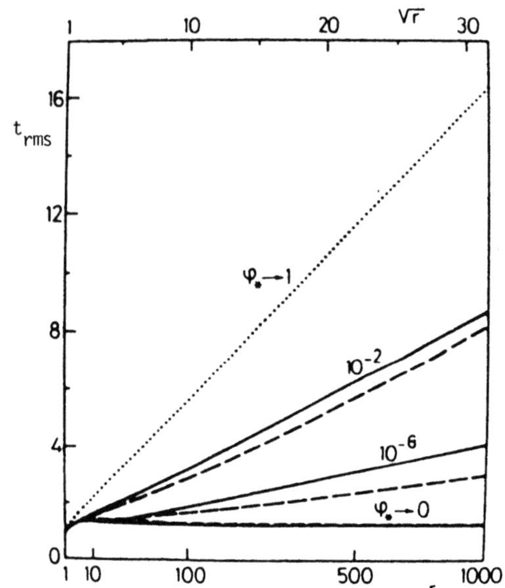

Fig. 10. The root-mean-square layer thickness t_{rms} as a function of the square root of chain length r for four values of polymer volume fraction ϕ_*[50]. Hexagonal lattice, $\chi_s = 1$. Full lines, $\chi = 0.5$; broken lines, $\chi = 0$; dotted line, bulk polymer

The average number of loops per chain, n_ℓ, increases with growing chain length and that of tails increases up to about 1.5. The average length of trains is about 4.3 independent of r, the loop length surpasses the train length and grows with increasing r, and the average tail size is approximately proportional to r. The fraction of segments in trains decreases with r, that in loops increases with r, and the tail fraction is independent of r. The fraction of trains of length s diminishes steeply with rising train length s, and the train segment distribution has a maximum at s = 4. If the segments in a train exceed 25 they contribute little to the number of trains or to the average number of train segments. The fraction of loops of length s decreases steeply with increasing loop length s, the corresponding fraction for tails decreases drastically with tail length s, and tails up to 800 segments give a significant contribution to the average number of tail segments. Although the calculation has been limited to r = 1,000 for the economy of computer time, the conclusion that tail segments overwhelmingly contribute to the thickness of the adsorbed layer is very important.

B.3.4 Root-Mean-Square Thicknesses of Loops and Tails

B.3.4.1 Root-Mean-Square Thickness of Loops

Experimentally, the root-mean-square thicknesses of loops and tails can be measured by ellipsometry. It is thus necessary to relate them to the average numbers of segments in loops and tails for deducing the conformation of an adsorbed polymer.

Hesselink[39,40] derived relations between the root-mean-square thickness and the loop or tail size. For the case in which no intrasegment interaction exists, he derived the segment distribution $\varrho_4(z)$ for a single loop of size i by considering all possible configurations of the chain that starts at the interface and returns at its end to the interface. His expression for $\varrho_4(z)$ reads

$$\varrho_4(z) = 12\,z(ib^2)^{-1}\exp(-6\,z^2/ib^2) \tag{B-103}$$

where b is the length of a polymer segment. Note that this distribution function is not exponential.
The loop size distribution defined by Eq. (B-7) can be rewritten

$$n_i = K' i^{-3/2} e^{i\lambda} \tag{B-104}$$

where K′ is a normalizing constant. For small λ, the average loop length $\langle i \rangle_\ell$ is given by $a(-\lambda)^{1/2}$. Hence, along with $n = \sum_i in_i$, Eq. (B-104) is rewritten

$$n_i = na\pi^{1/2}\langle i \rangle_\ell^{-1} i^{-3/2}\exp(-ia^2/\langle i \rangle_\ell^2) \tag{B-105}$$

where a is a numerical constant ($\simeq 0.7$). The normalized density distribution $\varrho_5(z)$ for loops of size $\langle i \rangle_\ell$ is represented by

$$\varrho_5(z) = n^{-1}\Sigma in_i\varrho_4(z) \tag{B-106}$$

Substitution of Eqs. (B-103) and B-105) into Eq. (B-106) gives

$$\varrho_5(z) = 2\,a\sqrt{6}\,(\langle i \rangle_\ell b)^{-1}\exp(-2az\sqrt{6}/\langle i \rangle_\ell b) \tag{B-107}$$

The mean-square thickness of loops for Hesselink's distribution function is thus given by

$$\langle z^2 \rangle = \int_0^\infty z^2\varrho_5(z)dz = \langle i \rangle_\ell^2 b^2/12\,a^2 \tag{B-108}$$

B.3.4.2 Root-Mean-Square Thickness of Tails

The segment density distribution $\varrho_6(z)$ for a single tail of size i is given by Hesselink[39, 40] as

$$\varrho_6(z) = 6\,(ib^2)^{-1} \int_z^{2z} \exp(-3\,t^2/2\,ib^2)dt \qquad \text{(B-109)}$$

Hesselink[51] assumed that the conformation of macromolecules adsorbed on the interface consists of one train and two tails. The number of tails with i segments, n(i), is calculated from the partition function for the tail-train conformation, and the normalized segment density distribution $\varrho_7(z)$ in tails is derived by methods analogous to those used in Hesselink's derivation of the segment density distribution in loops. The result is

$$\varrho_7(z) = 2\beta\,[\exp(-\beta z) - \exp(-2\beta z)] \qquad \text{(B-110)}$$

with

$$\beta = (3/\langle i \rangle_t b^2)^{1/2} \qquad \text{(B-111)}$$

where $\langle i \rangle_t$ is the average length of tails. Both $\varrho_6(z)$ and $\varrho_7(z)$ given by Eqs. (B-109) and (B-111) pass through a maximum at $z = 0.4\,(\langle i \rangle_t b^2)^{1/2}$. However, the distribution curve for a single tail falls off much *faster* than that for a set of tails. Therefore, the mean-square thickness of a single tail with size i is much less than that for a set of tails with an equal average size $\langle i \rangle_t$: Eq. (B- 109) gives $\langle z^2 \rangle = (7/18)ib^2$ whereas Eq. (B-110) gives $\langle z^2 \rangle = 7/6\,\langle i \rangle_t b^2$.

No analytical expression of the segment distribution for the loop-train-tail conformation has as yet been obtained.

B.3.5. Diffusion Equation Approach

The diffusion equation approach to polymer adsorption was first taken by Frisch[4]. He solved the diffusion equation for a polymer, adsorbed on a wall; this equation is subject to the boundary condition that the wall is completely reflective, i.e. the gradient of polymer concentration equals zero at the wall. However, DiMarzio[8] criticized that the correct boundary condition should be zero concentration at the wall. For a chain with its end attached to the interface he showed that the mean-square distance perpendicular to the surface is twice the value in the absence of the boundary. De Gennes[52] deduced an equation describing the adsorption of polymers at the theta point onto a planar adsorbing interface by applying the self-consistent mean field method. In his theory was no particular model considered for the conformation of adsorbed polymers. At the theta point, the probability distribution $G_n(\mathbf{r}, \mathbf{r}')$ for a polymer of n segments starting at $\mathbf{r}$ and ending at $\mathbf{r}'$ obeys the diffusion equation

$$\frac{1}{6}\,b^2\nabla^2 G_n(\mathbf{r}, \mathbf{r}') - \phi'(\mathbf{r})G_n(\mathbf{r}, \mathbf{r}') = \partial G_n(\mathbf{r}, \mathbf{r}')/\partial n \qquad \text{(B-112)}$$

where $\phi'(\mathbf{r})$ is the potential between a segment and the interface expressed in units of kT. If a square well potential is assumed for ϕ', Eq. (B-112) gives

$$G_n(\mathbf{r}, \mathbf{r}') \simeq 2\,k_0 \exp[-k(z+z')]\exp(nk_0{}^2b^2/6) \tag{B-113}$$

where z and z′ are distances corresponding to $\mathbf{r}$ and $\mathbf{r}'$, and the parameter k_0 is shown by Jones and Richmond[53] to be

$$k_0 = (6\phi'/b_2)^{1/2}\tan(6\phi' z_0{}^2/b^2)^{1/2} \tag{B-114}$$

with ϕ' and z_0 representing the depth and width of the square well potential.

From Eq. (B-113) the segment density distribution $\varrho(\mathbf{r})$ and the adsorption isotherm are calculated as follows:

$$\varrho(\mathbf{r}) = \sum_M \int dsdt G_M(s, \mathbf{r})G_{n-M}(\mathbf{r}, t)/\int dsdt G_n(s, t) \simeq n\,2\,k_0 \exp(-2\,k_0 z) \tag{B-115}$$

and

$$C = (2\varrho_b/k_0)\exp(nk_0{}^2b^2/6) \tag{B-116}$$

where C is the number of chains adsorbed per unit area and ϱ_b the concentration of chains per unit volume of bulk solution. The fraction of adsorbed segments p on the interface is given by

$$p \simeq \int_0^b \varrho(\mathbf{r})dz/\int_0^\infty \varrho(\mathbf{r})dz \simeq 1/(2\,k_0 b) \tag{B-117}$$

Jones and Richmond[53] extended de Gennes' derivation[52] to the case where excluded volume effects are present. For the adsorption from a good solvent, the probability distribution $G_n^\alpha(\mathbf{r}, \mathbf{r}')$ for an adsorbed polymer with n segments starting at $\mathbf{r}$ and ending at $\mathbf{r}'$ is governed by the diffusion equation

$$\frac{1}{6}\, b^2\nabla^2 G_n^\alpha(\mathbf{r}, \mathbf{r}') - [\nu(\mathbf{r}) + \phi'(\mathbf{r})]G_n^\alpha(\mathbf{r}, \mathbf{r}') = \partial G_n^\alpha(\mathbf{r}, \mathbf{r}')/\partial n \tag{B-118}$$

where $\nu(\mathbf{r})$ is defined by

$$\nu(\mathbf{r}) = \nu \sum_\alpha \sum_M \int dsdt G_M^\alpha(s, \mathbf{r})G_{n-M}^\alpha(\mathbf{r}, t)/\int dsdt G_n^\alpha(s, t) \tag{B-119}$$

The excluded volume ν is related to the parameter χ by

$$\nu = \left(\frac{1}{2} - \chi\right)v_s^2/v_1 \tag{B-120}$$

where v_s and v_1 denote the volumes of a segment and a solvent molecule, respectively. The mean thickness t of the adsorbed polymer layer is calculated to be

$$t = t_0(1 + 6\,nC\nu t_0/b^2) \tag{B-121}$$

where t_0 is the value of t at the theta point and given by $1/(2\,k_0)$. If the ratio t/t_0 is defined as the expansion factor α_t of the adsorbed polymer layer, Eq. (B-121) is rewritten

$$\alpha_t - 1 = 6\,nC\nu t_0/b^2 \tag{B-122}$$

Moreover, the number of polymers adsorbed per unit area, C, is given by

$$C \simeq C^0(1 - k_0 n^2 \nu C^0) \tag{B-123}$$

where C^0 is the value of C at the theta point.

B.3.6 The Scaling Theory

De Gennes[24, 54] investigated the adsorption of a flexible polymer on a flat surface from good solvents using the scaling theory. Three different regimes, i.e. dilute, semidilute, and plateau regions, were considered.

1) The dilute region
For a single chain of n segments confined in the adsorbed polymer layer with thickness D, the chemical potential μ is given by

$$\mu - \mu_0 = n(b/D)^{5/3} - \delta(bn/D) + \mu_{trans} \tag{B-124}$$

where μ_0 is the chemical potential of a free chain, $n(b/D)^{5/3}$ the confinement energy for a self-excluded chain, $\delta(bn/D)$ an effective adsorption energy, and μ_{trans} is expressed by

$$\mu_{trans} = \ln(\Gamma/n) + \text{const} \tag{B-125}$$

with Γ denoting the adsorbed amount (segments/cm^2). Differentiating Eq. (B-124) with respect to D, we obtain for the optimum value of D

$$D_{\text{single chain}} = b\delta^{-3/2} \tag{B-126}$$

D is independent of n and increases with decreasing δ. Eq. (B-124) is rewritten using Eq. (B-126) as follows:

$$\mu = \mu_0 - n\delta^{5/2} + \mu_{trans} \tag{B-127}$$

If Eq. (B-127) is equated to the chemical potential μ_b in the bulk solution with concentration C_b, we get

$$\mu_b \cong \ln(C_b/n) \tag{B-128}$$

Hence, the adsorption isotherm is given by

$$\gamma \cong b^3 C_b \exp(n\delta^{5/2}) \tag{B-129}$$

where γ is equal to Γb^2, which is a dimensionless surface polymer concentration.

2) The plateau region
In the plateau region many polymer chains are adsorbed and give rise to an adsorbed layer with thickness D. The chemical potential μ in this region is expressed by

$$\mu - \mu_0 \cong n(b/D)^{5/3} - n\delta(b/D) + n(Cb^3)^{5/4} + \mu_{trans} \tag{B-130}$$

where $n(Cb^3)^{5/4}$ is the contribution from the osmotic pressure in the adsorbed layer and equal to $n(\Gamma b^2/D)^{5/4}$. The confinement energy $n(b/D)^{5/3}$ is negligibly small in comparison with the osmotic term. Under these conditions, differentiation of Eq. (B-130) with respect to D gives for the optimum D

$$D \cong b(\gamma^5/\delta^4) \tag{B-131}$$

which allows Eq. (B-130) to be rewritten

$$\mu = \mu_0 - n(\delta/\gamma)^5 + \mu_{trans} \tag{B-132}$$

If this is equated to μ_b, we get for the adsorption isotherm

$$\gamma \cong \delta n^{1/5} \ell^{-1/5} \tag{B-133}$$

where ℓ is defined as $\ln(\gamma/b^3 C_b)$. Eq. (B-133) shows a very weak dependence of γ on n. The thickness D is obtained by substituting Eq. (B-133) into Eq. (B-131):

$$D(C_b) = bn\delta\ell^{-1} \tag{B-134}$$

which indicates that in the plateau region D is proportional to n since ℓ is nearly constant. However, $D/nb \sim \delta$ is so small that the chain is far from being in a stretched conformation.

3) The semidilute region
In the semidilute region, the repulsion between adsorbed chains is dominated by the osmotic term. Hence, the chemical potential μ is represented by

$$\mu - \mu_0 = n(b/D)^{5/3} + n(b/D)(\gamma^2 - \delta) + \mu_{trans} \tag{B-135}$$

where $n(b/D)\gamma^2$ corresponds to the osmotic factor in two dimensions. After differentiating Eq. (B-135) with respect to D we find for the optimum D

$$D/b \cong (\delta - \gamma^2)^{-3/2} \tag{B-136}$$

which allows Eq. (B-135) to be rewritten

$$\mu \cong -n(\delta - \gamma^2)^{5/2} + \mu_{trans} \tag{B-137}$$

The adsorption isotherm is obtained, by equating Eq. (B-137) to the chemical potential μ_b of the bulk solution, as

$$\gamma^2 \cong \delta - (\ell/n)^{2/5} \tag{B-138}$$

From Eqs. (B-136) and (B-138) D is found to be

$$D \cong b(n/\ell)^{3/5} \tag{B-139}$$

B.4 Theories of the Adsorption of Polyelectrolytes

B.4.1 The Theory of Frisch and Stillinger

Frisch and Stillinger[55] were the first to derive the isotherm for a rigid polyelectrolyte adsorbed on a dielectric surface. Assuming that a rod-like polyion is oriented parallel to the surface, they calculated the electrostatic free energy needed to bring the polyion to the surface against the potential of an image force, formed by the dielectric surface, and combined it with the adsorption energy to derive an adsorption isotherm. It was shown that if the dielectric constant of the surface is much larger, than that of the solution, adsorption of rod-like polyions is enhanced by increasing the concentration of an added simple salt, but the fraction of charged groups of the polyelectrolyte adsorbed on the surface is relatively low. Their theory has little practical interest, since most polyelectrolytes are flexible.

B.4.2 The Theory of Hesselink

Hesselink[23] attempted to calculate adsorption isotherms for flexible polyelectrolytes. He assumed that, when adsorbed on a surface, a flexible polyelectrolyte takes a conformation consisting of one train and one tail. The theoretical treatment of Hoeve et al.[41] (cf. B.3.1) for non-ionic polymers was extended by taking into account the change in electrical free energy that occurs when the polyelectrolyte is brought from the solution onto the interface. The partition function Q for a system of N polyelectrolytes each consisting of n units, in which N_a polyions are adsorbed on the surface of area S and $N_f(N_f = N - N_a)$ polyions remain in the bulk solution of volume V, is then represented by

$$Q = \sum_{N_a=1}^{N} \exp(-\Delta F^e/kT) \prod_i (\omega_i \nu_{n-i})^{n_i^t}/n_i^t!/[(S\,\delta)^{N_a} V^{N_f}/N_f!] \tag{B-140}$$

The quantity ΔF^e is the electrical free energy of adsorption of N_a polyions, ω_i and ν_{n-i} are the partition functions for a tail of i units and that for a train of $j(=n-i)$ units, n_i^t is the number of tails of i units, and δ the first layer thickness. The partition function ω_i is given by

$$\omega_i = c i^{-1/2} \tag{B-141}$$

where c is the flexibility parameter of polyelectrolyte. On the other hand, the partition function ν_{n-i} is given by

$$\nu_{n-i} = \sigma^{n-i} \tag{B-142}$$

when σ is the internal partition function of a segment adsorbed on the surface relative to that in the bulk solution. Furthermore, we have the conditions:

$$\Sigma n_i^t = N_a \tag{B-143}$$

$$N_a + N_f = N \tag{B-144}$$

Now, we can express ΔF^e as

$$\Delta F^e = \Delta F_s^e + F_t^e - F_c^e \tag{B-145}$$

where ΔF_s^e is the increase in the electrical free energy of the surface, F_t^e the electrical free energy of the layer of tails, and F_c^e the electrical free energy of N_a random coils in solution. The term F_c^e is given, according to Hermans and Overbeek[56], by

$$F_c^e = N_a 2\pi \varrho_0^2 s_0^3/\varepsilon \varkappa^2 \tag{B-146}$$

where ϱ_0 is the charge density inside the coil, s_0^3 the volume of the polymer coil in the bulk solution, ε the dielectric constant of the bulk solution, and $\varkappa^{-1}$ the Debye length. The free energies ΔF_s^e and F_t^e can be obtained by solving the Poisson equation for the electrical potentials of the surface and the tails, yielding

$$\Delta F_s^e = (2\pi S/\varepsilon\varkappa)(2\sigma_0\sigma_1 + \sigma_0\varrho_v/\varkappa + \sigma_1^2\sigma_1\varrho_v\varkappa) \tag{B-147}$$

$$F_t^e = (2\pi S/\varepsilon\varkappa^2)\varrho_v(\sigma_0 + \sigma_1 + \varrho_v) \tag{B-148}$$

Here σ_1 is the density of surface charges generated by the adsorbed train units, σ_0 the original density of surface charge, ϱ_v the volume charge density in the adsorbed layer of dangling tails, and s the adsorbed thickness normal to the surface. Substituting Eqs. (B-141), (B-143), (B-146), (B-147), and (B-148) into Eq. (B-140) and maximizing the partition function Q with respect to n_i^t, N_a, and N_f, we obtain

$$n_i^t = (S\delta/V)ci^{-1/2}\sigma^{n-i}N_f\exp[-(n-i)(4\pi\alpha e/\varepsilon kT\varkappa)(\alpha_0 + \alpha e\Gamma)] \tag{B-149}$$

with αe denoting the number of charges per segment and Γ the adsorbed amount. Since Γ is equal to $(nN_a - \Sigma i n_i^t)/S$ or $(n\Sigma n_i^t - \Sigma i n_i^t)/S$, the adsorption isotherm is derived from Eq. (B-149) as follows:

$$nN_f/V = \Gamma\exp\{-n[\ln\sigma - U(\sigma_0 + \alpha e\Gamma)]\}/c\delta(S_{-1/2} - n^{-1}S_{1/2}) \tag{B-150}$$

$$\equiv \Gamma\exp(-nE)/c\delta(S_{-1/2} - n^{-1}S_{1/2}) \tag{B-151}$$

where

$$U = 4\pi\alpha e/(\varepsilon kT\varkappa) \qquad \text{(B-152)}$$

$$S_a = \sum_{i=1}^{n} i^a \exp\{-i[\ln\sigma - U(\sigma_0 + \alpha e\Gamma)]\} \qquad \text{(B-153)}$$

Equations (B-150) and B-151) show that adsorption of a flexible polyelectrolyte reaches saturation when the electrical term is equal to $\ln\sigma$. The value of Γ at the saturation is given by

$$\Gamma_\infty = \varepsilon kT\varkappa \ln\sigma(4\pi\alpha^2e^2)^{-1} - \alpha_0/\alpha_e \qquad \text{(B-154)}$$

It is seen that Γ_∞ is independent of the polymer molecular weight and that Γ_∞ increases with rising square root of the added salt concentration in the bulk solution when $\sigma_0 = 0$. The polymer concentration in the bulk solution at the saturation is evaluated from Eq. (B-150) to be

$$(nN_f/V)_\infty \approx 3\,\varepsilon kT\varkappa \ln\sigma/(16\pi\alpha^2e^2\,c\delta n^{1/2}) \qquad \text{(B-155)}$$

Thus, $(nN_f/V)_\infty$ is proportional to the reciprocal square root of the molecular weight. The fraction f^t of the segments in the tail is given by

$$f^t = \Sigma\, i\, n_i^t/nN_a = \Sigma\, i\, n_i^t/n\,\Sigma\, n_i^t = n^{-1}S_{1/2}/S_{-1/2} \qquad \text{(B-156)}$$

The last term follows Eqs. (B-151) and (B-153). Numerical calculations show that, at the saturation, f^t is independent of n as well as other variables. It appears that this theory of Hesselink seems to explain qualitatively the observed phenomena, but the conformational model of one tail and one train is too simple to be accepted; it is unreasonable to ignore the formation of loops.

In his second paper, Hesselink[57] developed a new theory for the adsorption of polyelectrolyte by assuming loop-train conformations but neglecting tail formation. The partition function for N_a polyelectrolytes adsorbed on the surface with an area S from the solution of volume V is still given by Eq. (B-21), but ΔF is now the sum of the following five contributions.

1) ΔF_1 is due to the configurational change on adsorption and given, according to Hoeve[42], by

$$\Delta F_1/kT = -\,nN_a[up - \lambda(1-p)] \qquad \text{(B-157)}$$

where λ is the Lagrangian multiplier, and u is given by

$$u = \ln\sigma(1 + S_{-3/2}) \qquad \text{(B-158)}$$

The term $S_{-3/2}$ is already formulated in the theory of Hoeve et al.[41]

2) ΔF_2 is associated with the interactions of adsorbed trains themselves and with the solvent in the first adsorbed layer and given by

$$\Delta F_2/kT = (S\delta/v_1)[(1-\nu_2(0))\ln(1-\nu_2(0)) + \nu_2(0)(1-\chi\nu_2(0))] \qquad \text{(B-159)}$$

3) ΔF_3 is due to non-ionic interactions between dangling loops and the solvent and expressed by

$$\Delta F_3/kT = (S\,s/v_1)\left(\frac{1}{2} - \chi\right) v_2^2 \qquad \text{(B-160)}$$

where v_2 is the volume fraction of segments in the loop layer and s the thickness of the adsorbed layer. The parameter s is related to the volume V_0 of a polyelectrolyte coil in the bulk solution by

$$s = (1 - p)\, V_0 N_a/S \qquad \text{(B-161)}$$

where $1 - p$ is the fraction of segments in the loops.

4) ΔF_4 is the change in the free energy of the electrical double layer accompanying the adsorption of charged trains on the charged surface, and if the Debye-Hüchel approximation is applied, it is given by

$$\Delta F_4 = 2\pi S/\varepsilon\varkappa(\sigma_t^2 - \sigma_0^2 + 2\sigma_0\varrho_{pol}/\varkappa) \qquad \text{(B-162)}$$

where ε is the dielectric constant of the solution part in the double layer, $\varkappa^{-1}$ the Debye length, σ_0 the charge density of the surface, ϱ_{pol} the volume charge density due to loops. The total surface charge density σ_t is given by

$$\sigma_t = \sigma_0 + \tau\alpha\, enpN_a/S \qquad \text{(B-163)}$$

with $\tau\alpha e$ denoting the charge per adsorbed segment, where e is the protonic charge and α the degree of dissociation of the polyelectrolyte. Furthermore, $\tau = -1$ for an anionic polymer and $\tau = 1$ for a cationic polymer.

5) ΔF_5 is due to dipole and/or polarizability effects which arise when an uncharged segment is adsorbed on the charged surface. It is represented by

$$\Delta F_5/kT = npN_aB\sigma_t^2 \qquad \text{(B-164)}$$

where B is a constant depending on the system.

Substituting Eqs. (B-157), (B-159), (B-160), (B-162), and (B-164) into Eq. (B-156) and maximizing the partition function Q with respect to p and N_a, we get

$$-\lambda = u + \ln(1 + v_2(0)) + 2\chi v_2(0) + (1/2 - \chi)v_2 - \tau\alpha\varkappa\sigma_t/2\,ne - b'\sigma_t(3\sigma_t - 2\sigma_0) \qquad \text{(B-165)}$$

$$N_a/S\delta = (N_f/V)\exp[n\{u + \ln(1 - v_2(0)) + 2\chi v_2(0) - \tau\alpha\varkappa\sigma_t/2\,ne - b'\sigma_t(3\sigma_t - 2\sigma_0)\}] \qquad \text{(B-166)}$$

The adsorption isotherm given by the last equation reduces to that for non-ionic polymers when $\sigma_t = 0$.

Some interesting conclusions can be drawn from this theoretical treatment of the adsorption of polyelectrolyte. The calculated isotherms are of the high-affinity type as is usually observed. The adsorbance drastically increases and the fraction of adsorbed segments steeply decreases at high surface coverage. At low adsorption energy the adsorption of an anionic polyelectrolyte on a positively charged surface is in general higher than that on a neutral or negatively charged surface. The adsorbance increases with increasing salt concentration. This is expected because a rise in the salt concentration reduces the electrostatic repulsion between adsorbed polyelectrolytes. As the degree of dissociation is increased the adsorbance usually decreases due to an increase in the volume of the polyelectrolyte coil. In contrast to the theory using the one-tail and one-trail model, the adsorbance at the plateau increases with rising n, the number of segments in the polyelectrolyte, according to the relation

$$\ln(N_a/S) = a' \ln n + b' \tag{B-167}$$

where a′ and b′ are constants. The calculated thickness of the adsorbed layer given by Eq. (B-161) is overestimated, since the segment density in the loop layer is assumed to be equal to that in the bulk solution. Hesselink suggested that the calculated thickness should be divided by a factor of 8 in order to bring it to the right order of magnitude.

B.5 Summary

For the conformation of adsorbed polymer chains, the loop-train model and/or the loop-train-tail model are now widely accepted. In formulating the theory, segment-segment, segment-surface, and segment-solvent interactions are taken into consideration. Regardless of the model, all the existing theories predict that at high adsorption energy, the conformation of adsorbed polymer chains is flat and the fraction of adsorbed segments is large. The conformation of adsorbed polymer chains is controlled by the segment distribution in the adsorbed layer. The root-mean-square thickness of this layer may be used to interpret the experimental data obtained, for examble, by ellipsometry (cf. Sect. C.2.2). Theories based on the loop-train model indicate that this average thickness at the theta point is proportional to the square root of the polymer molecular weight. However, the same molecular weight dependence is also derived from the loop-train-tail model. Furthermore, this average thickness is predominantly determined by tails rather than by loops. Existing theories allow other important quantities such as the adsorbed amount, the fraction of adsorbed segments, and the surface coverage to be calculated, but the results from different theories are not very different from one another. The diffusion equation approach and the scaling theory lead to very simple results.

Existing theories of the adsorption of polyelectrolyte allow effects of the polymer charge density, the surface charge density, and the ionic strength on the adsorption behavior to be predicted. The predicted adsorption behavior resembles that of nonionic polymers if the ionic strength is high or the polymer charge density is very low.

C. Experimental Techniques

As has been depicted in Fig. 1, various conformations are possible for adsorbed polymers, depending on polymer-polymer, polymer-solvent, and polymer-interface interactions and the flexibility of polymers. To determine experimentally the conformation of adsorbed polymers only adsorption isotherm data are insufficient. The average thickness of the adsorbed polymer layer, the segment density distribution in this layer, the fraction of adsorbed segments, and the fraction of surface sites occupied by adsorbed segments must be measured. Recently, several unique techniques have become available to measure these quantities.

C.1 Adsorption Isotherms

Adsorption isotherms are readily determined by measuring polymer concentrations in the bulk solutions before and after adsorption equilibrium has been attained. However, the time required to reach equilibrium is often considerable.

Measured isotherms are usually of the high-affinity type for which reliable and accurate data can be obtained only in the plateau region of the isotherm. The initial rising part of the isotherm is often difficult to measure accurately, because we have to determine trace amounts of polymer

Cohen-Stuart et al.[58] demonstrated that the molecular weight distribution affects the shape of adsorption isotherms. In fact, often observed round-shaped isotherms are attributed to a broad molecular weight distribution.

C.2 Thickness of the Adsorbed Layer

C.2.1 Hydrodynamic Methods

The most convenient of these methods is viscosity measurement of a liquid in which particles coated with a polymer are dispersed, or measurement of the flow rate of a liquid through a capillary coated with a polymer. Measurement of diffusion coefficients by photon correlation spectroscopy as well as measurement of sedimentation velocity have also been used. Hydrodynamically estimated thicknesses are usually considered to represent the correct thicknesses of the adsorbed polymer layers, but it is worth noting that recent theoretical calculations[52, 59] have shown that the hydrodynamic thickness is much greater than the average thickness of loops.

C.2.2 Ellipsometry

Ellipsometry[27, 60–62] is based on the principle that light undergoes a change in polarizability when it is reflected at a surface. The refractive index of the surface and the reflection coefficient of a system can be calculated from the change in the phase retardation Δ and the change in the amplitude ratio tan ψ. Adsorption of a polymer on a surface gives rise to additional changes in Δ and tan ψ, which allow the thickness and refractive index of the adsorbed layer to be determined. However, since the thickness so determined concerns the thickness of a hypothetical homogeneous layer, the root-mean-square thickness must be estimated therefrom by making appropriate assumptions for the distribution of segments in the adsorbed layer. This root-mean-square thickness may be compared with the root-mean-square thickness for loops or tails predicted by theory. However, the segment distribution itself cannot be measured directly. The refractive index gives the average segment density in the adsorbed layer, and its product with the average thickness of the adsorbed layer gives the adsorbance. The advantage of ellipsometry allows in situ measurements of the refractive index and the thickness of the adsorbed layer to be made, although its applicability is limited to the adsorption onto a flat, smooth, reflective surface, i.e. a metallic or mirror surface.

C.2.3 The ATR Method

The attenuated total reflection (ATR) method measures the reflection coefficients of vertically and horizontally polarized light reflected from a polymer layer adsorbed on a transparent surface[63]. These coefficients allow the thickness of the adsorbed layer and the polymer concentration in it to be determined.

In principle, the ATR method would provide information about the segment distribution in the adsorbed layer if light could penetrate in different depths into the layer, but this possibility still remains untested.

C.3 Fraction of Adsorbed Segments and Fraction of Occupied Surface Sites

The infrared band of a particular group in a polymer shifts when some groups of the polymer are adsorbed onto active sites, e.g. silanol groups on a silica surface[25]. This phenomenon has been used to measure the fraction of adsorbed polymer segments. The fraction of the surface sites occupied by adsorbed polymer segments can also be determined from the frequency shift of IR band caused by the interaction between functional groups and an active site.

When spin-labeled species are chemically attached to a polymer at random, the difference in mobility between the labeled segments adsorbed (trains) and unadsorbed (loops or tails) gives rise to a variation in the magnetic relaxation time. This difference allows an estimation of the fraction of adsorbed segments if the signals from the adsorbed and unadsorbed labeled segments can be separated. However, the applicability of this electron paramagnetic resonance (EPR) method[64] to polymer adsorption has two limilations. One is that the introduction of the spin-labeled species sometimes affects the adsorption behavior. The other is that the mobility of unadsorbed segments is often disturbed by adjacent adsorbed segments. The latter is notable for short loops, leading to mistaking such loops as trains. Thus, the EPR method may estimate the fraction of adsorbed segments over the values obtained by the IR method.

C.4 Heat of Adsorption

Calorimetry[65] is the only direct method concerning the energetics of adsorption processes. The heat of adsorption should be generated from the difference between surface-segment and surface-solvent interactions, i.e. the χ_s parameter, and from the difference among segment-segment, segment-solvent, and solvent-solvent interactions. No relationship between the heat of adsorption and these parameters has as yet been established.

D. Experimental Results

D.1 Adsorption of Flexible Non-Ionic Polymers

D.1.1 Ellipsometric Studies

D.1.1.1 Adsorption at the Theta Point

Ellipsometry determines a certain average thickness t_h of the adsorbed layer. However, what is important for the evaluation of polymer conformations in this layer is the root-mean square thickness t_{rms}. Hence, it is necessary to find a way of relating t_{rms} to t_h. McCrackin and Colson[66] studied this problem for several distributions of segments and found $t_{rms} = t_h/1.5$ for the exponential distribution and $t_{rms} = t_h/1.74$ for the Gaussian distribution. Takahashi et al.[67] showed that $t_{rms} = t_h/1.63$ for the one-train and two-tail model (see Eqs. (B-110) and (B-111)).

Stromberg et al.[68] applied ellipsometry to the adsorption of polystyrene samples with a narrow M-distribution on a chrome plate from cyclohexane solutions at 35 °C (the theta condition for polystyrene) and found that t_{rms} was approximately proportional to the square root of molecular weight, M up to 1.8×10^6. These values of t_{rms} were compared with the root-mean-square end-to-end distances calculated for the chains which are attached at one end to either a reflecting or an absorbing wall. They fell between the computed values for these two walls, though somewhat closer to the values for the absorbing wall.

Peyser and Stromberg[63] used the ATR method to measure the thickness of a polystyrene layer adsorbed on a quartz surface from cyclohexane solution at 35 °C and compared it with the thickness obtained by ellipsometry. Good agreement was observed although the ellipsometric measurements were made for the chrome plate.

Gebhard and Killmann[69] reported an ellipsometric study of the adsorption of polystyrene onto various metal surfaces from theta solvents, M ranging from 76×10^3 to 340×10^3. A proportionality between t_{rms} and $M^{1/2}$ was also observed, and the adsorbance was found to increase with rising M.

Recently, Takahashi et al.[70] measured ellipsometrically the thicknesses of the polystyrene layers adsorbed on a chrome plate from cyclohexane at 35 °C over a very wide range of M from 10×10^3 to 13.4×10^6. The resulting adsorption isotherms are shown in Fig. 11. Initially, the adsorption isotherms rise sharply with increasing bulk polymer concentration and reach a wide plateau region. The plots of t_{rms} against bulk polymer

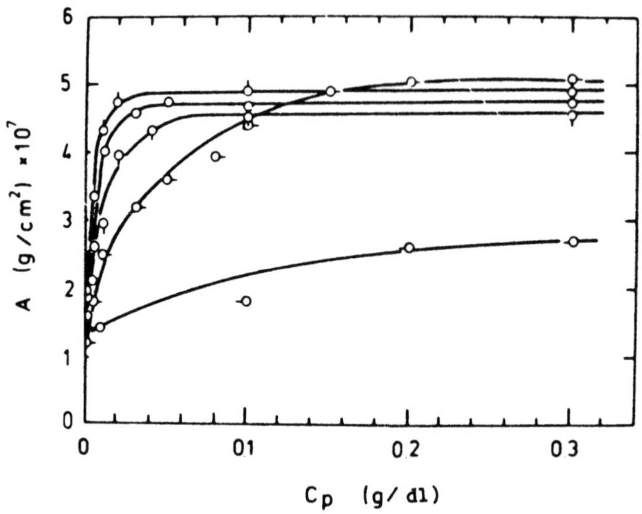

Fig. 11. Adsorption isotherms of polystyrene[70]: (-O) $M_w = 110 \times 10^3$; (O-) $M_w = 670 \times 10^3$; (O̧) $M_w = 2.42 \times 10^6$; (Ȯ) $M_w = 7.62 \times 10^6$; (O) $M_w = 13.4 \times 10^6$

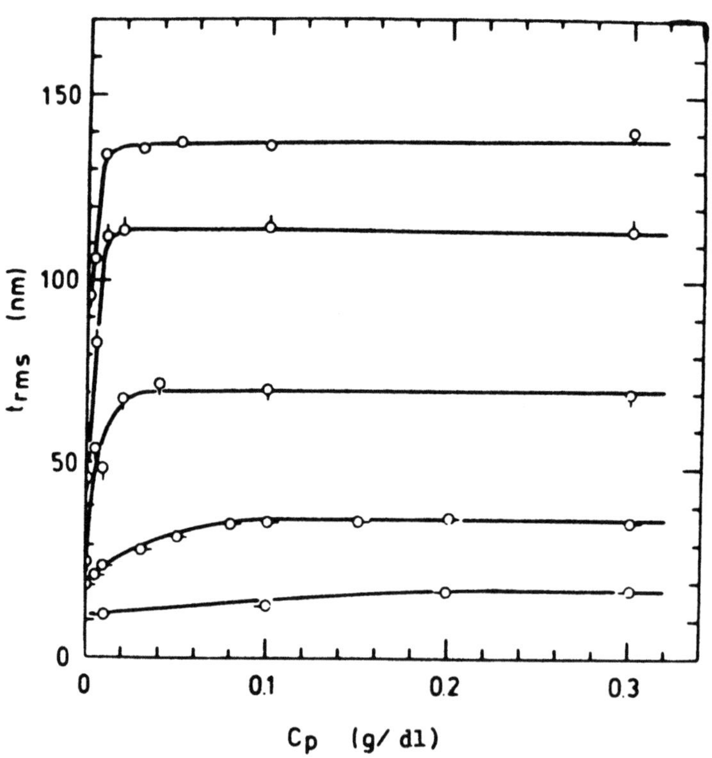

Fig. 12. Root-mean-square thickness t_{rms} of the adsorbed polymer layer vs. polymer concentration[70]. Symbols are the same as in Fig. 11

concentration are shown in Fig. 12. The thickness also rises quite steeply with increasing bulk polymer concentration and then levels off at a constant value. At very low bulk polymer concentration, both t_{rms} and the adsorbance are low which indicates that the polymer assumes a flattened conformation consisting of small loops or tails and a large number of train segments. A further increase of the bulk polymer concentration causes a desorption of train segments and makes adsorption sites available for other polystyrene molecules, leading to a thicker adsorbed layer. Thus, the polymer is expanded in the direction normal to the surface. The values of t_{rms} in the plateau region are proportional to $M^{1/2}$ as shown in Fig. 13. The average polymer concentration in the adsorbed layer first increases up to $M = 500 \times 10^3$, reaches a maximum, and ultimately decreases linearly as M is increased. The slope of the decreasing portion in the double-logarithmic plot is -0.5. These concentrations of adsorbed polymers are about four times as large as that of a single isolated random coil. The adsorbance A_p in the plateau region rises linearly with increasing M up to 500×10^3; above this molecular weight it is independent of M, as illustrated in Fig. 14.

Perkel and Ullman[71)] suggested the following empirical relation between A_p and M:

$$A_p = k'M^a \qquad \text{(D-1)}$$

where k' and a are constants. The constant a bears some relation to the polymer conformation; thus, $a = 1$ for single-point attachment, $a = 1/2$ for a random coil, and $a = 0$ for all polymer segments located on the interface. However, the comparison of our experimental data and theories, which is presented below, shows that it is misleading to infer polymer conformations only from the value of a.

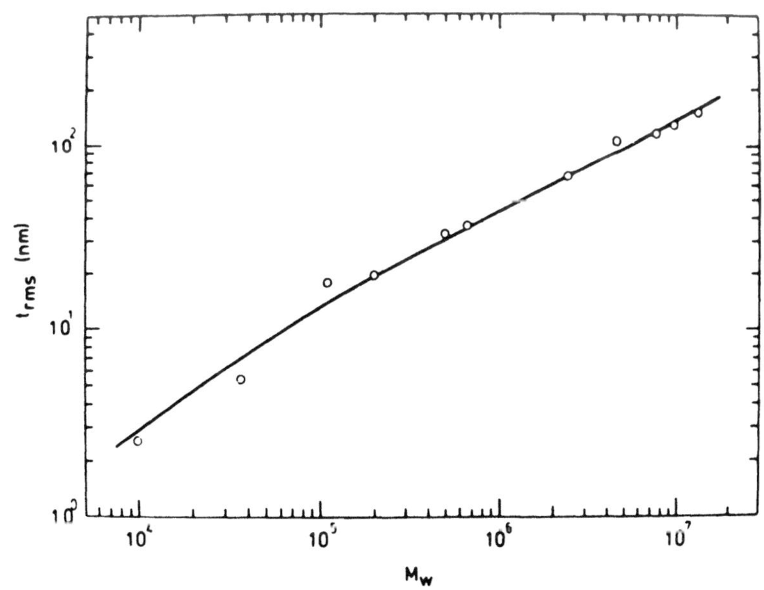

Fig. 13. Root-mean-square thickness t_{rms} of the adsorbed polymer layer at the bulk polymer concentration 0.3 g/dl vs. molecular weight[70)]

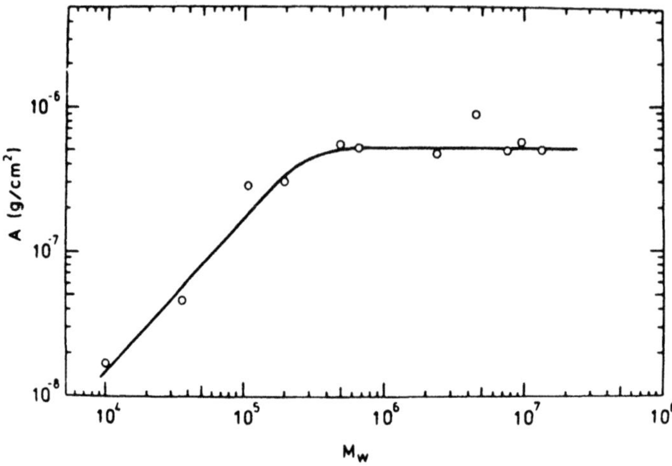

Fig. 14. Adsorbance A at the bulk polymer concentration 0.3 g/dl vs. molecular weight[70)]

We have compared our experimental results with the theories of Hoeve[42)], Silberberg[47)], and Scheutjens and Fleer[48)]. In doing so, the following values were selected for various parameters. In the theory of Silberberg, the conformational parameter $\gamma_s\gamma_B = 0.1$, the segment-surface interaction parameter $\chi_s = \infty$, and the Flory-Huggins parameter $\chi = 0.5$. In the theory of Scheutjens and Fleer, $\chi_s = 5$ and $\chi = 0.5$. Both experimental adsorption isotherms and thicknesses of the adsorbed layers were in poor agreement with the theory of Hoeve. As seen from Figs. 15 and 16, the molecular weight dependence of the adsorbance is in excellent agreement with Silberberg's theory, but the theoretical average loop size is too small to be consistent with the large measured value of

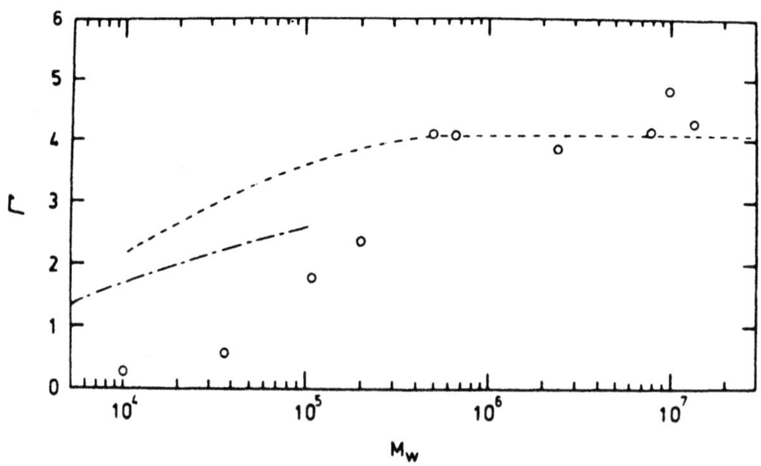

Fig. 15. Asorbed amount Γ as a function of molecular weight[70)]. Dependence predicted by theory for a θ-solvent ($\chi = 0.5$); (----) by the theory of Silberberg[47)] for $\gamma_s\gamma_B = 0.1$, $\chi_s = \infty$, and $\phi^* = 3 \times 10^{-3}$; (——) by the theory of Scheutjens and Fleer[48)] for $\chi_s = 5$ and $\phi_* = 10^{-3}$

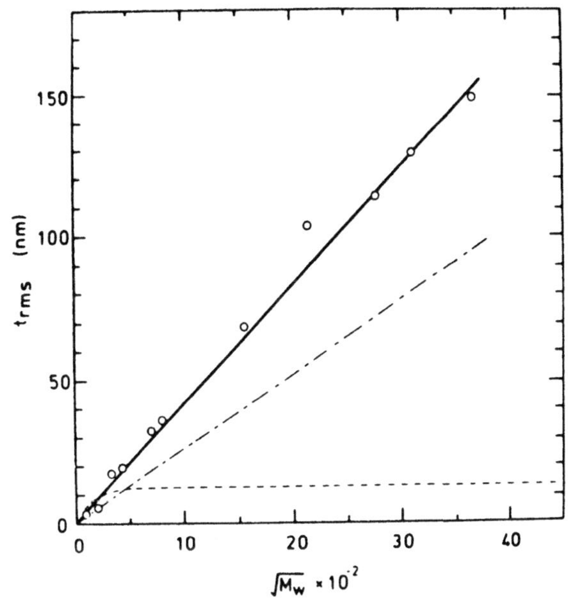

Fig. 16. Root-mean-square thickness t_{rms} of the adsorbed layer as a function of the square root of molecular weight[70] (----) by the theory of Silberberg[47] for $\gamma_s\gamma_B = 0.1$, $\chi_s = \infty$, and $\phi^* = 3 \times 10^{-3}$; (———) by the theory of Scheutjens and Fleer[48]

t_{rms}. Scheutjens and Fleer[48] showed that about 15% of the segments in the adsorbed layer may be present in one tail. Furthermore, they showed that t_{rms} increases in proportion to $M^{1/2}$. Therefore, t_{rms} may be expressed by

$$t_{rms} = \left(\frac{15}{100} n\right)^{1/2} b \qquad \text{(D-2)}$$

where n is the degree of polymerization and b the effective bond length (0.69 nm)[72]. Equation (D-2) gives t_{rms} close to the observed t_{rms}, as illustrated by the solid line in Fig. 16. Thus, we see that the portion of tails consisting of 15 to 20% of the total segments in the adsorbed layer dominates the thickness of this layer and that the polystyrene molecule adsorbed onto the chrome plate assumes a loop-train-tail conformation.

For the adsorption of polystyrene onto a mercury surface[73] from cyclohexane solutions at 35 °C, the thickness of the adsorbed layer was independent of molecular weight. This may be attributed partly to a higher adsorption energy and partly to the fluidity of the mercury surface.

Gebhard and Killmann[69] measured the thickness of the adsorbed layer and the adsorbance for poly(methyl methacrylate) adsorbed onto a high-vaccum vapored chromium plate and a platinum plate from theta solvents: 1-chlorobutane and acetonitrile. Both quantities increased with rising molecular weight, but the average concentration of adsorbed poly(methyl methacrylate) decreased with molecular weight. The thickness of the adsorbed layer was not proportional to the square root of M.

D.1.1.2 Adsorption from Good Solvents

Kawaguchi and Takahashi[74] studied the adsorption of polystyrene onto a chrommium plate from cyclohexane at 40 and 45 °C and found that the thickness of the adsorbed layer increased and the adsorbance decreased with rising temperature. The slope of a double-logarithmic plot of the thickness of the adsorbed layer versus the molecular weight increased with temperature. Both the adsorbance and the thickness of the adsorbed layer changed reversibly upon a cyclic change in temperature. This indicates that the conformation of the adsorbed polymer changed reversibly. The expansion factor α_t defined as the thickness of the adsorbed layer in a good solvent (T) relative to that at the theta point (Θ) was compared with the theoretical predictions of Hoeve[44, 45] and Jones and Richmond[53].

In the theory of Hoeve based on the loop-train model, α_t is correlated with the excluded volume derived from Eq. (B-34) by

$$\alpha_t^3 - \alpha_t = 4\pi^{1/2}\beta_1 n N_p (V_1/N_A)\,\psi_1(1 - \Theta/T)/15.37 \qquad \text{(D-3)}$$

where N_p is the number of adsorbed polymers per unit area, V_1 the molar volume of solvent, N_A the Avogadro constant, ψ_1 the parameter for the entropy of dilution of polymer with solvent, and $\beta_1 = (3n/2\langle r_n\rangle^2)^{1/2}$, with $\langle r_n\rangle^2$ denoting the unperturbed mean-square end-to-end distance of a random coil whose degree of polymerization is n. By introducing $\beta_1^{-1} = 0.56$ (nm)[72], $(V_1/N_A) = 0.185$ (nm^3)[75], and $\psi_1 = 0.13$[76], Eq. (D-3) is written

$$(\alpha_t^3 - \alpha_t)/n = 1.94 \times 10^{-16} N_p(1 - \Theta/T) \qquad \text{(D-4)}$$

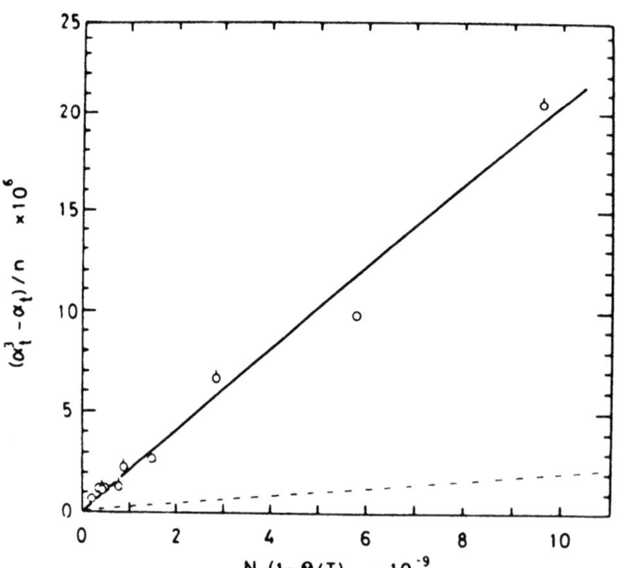

Fig. 17. Plot of $(\alpha_t^3 - \alpha_t)/n$ vs. $N_p(1 - \Theta/T)$[74]. The dotted line is calculated from Eq. (D-4); (○), 40 °C; (ơ), 45 °C

Figure 17 compares experimental data with Eq. (D-4). The data points behave as predicted by Eq. (D-4). However, the slope of the solid line is 2.0×10^{-15}, which is one order of magnitude larger than that of the dashed line predicted by Eq. (D-4).

The theory of Jones and Richmond based on the self-consistent mean field theory predicts Eq. (B-122), which may be rewritten

$$(\alpha_t - 1) = 6\,nN_p t_{rms(\Theta)}\, \psi_1 (V_1/N_A)(1 - \Theta/T)/b^2 \qquad \text{(D-5)}$$

where $t_{rms(\Theta)}$ is the value of t_{rms} at the theta temperature. Taking $(V_1/N_A) = 0.185$ (nm^3), $\psi_1 = 0.13$, and $b = 0.69$ nm[72], we can express Eq. (D-5) as

$$(\alpha_t - 1)/n = 3.08 \times 10^{-8} N_p t_{rms(\Theta)} (1 - \Theta/T) \qquad \text{(D-6)}$$

The values of $(\alpha_t - 1)/n$ are plotted against $N_p t_{rms(\Theta)} (1 - \Theta/T)$ in Fig. 18. The predicted linear relation is not obeyed and the measured values of $(\alpha_t - 1)/n$ are far smaller than the theoretical ones. Thus, wc find that the excluded volume effect formulated by Jones and Richmond is entirely irrelevant.

The smaller expansion factor predicted by the theory of Hoeve originates from neglecting the tail portions of the adsorbed polymer chain, while the larger expansion factor predicted by Jones and Richmond is due to their failure of correctly evaluating the elastic free energy, as has been pointed out by Kawaguchi and Takahashi[74].

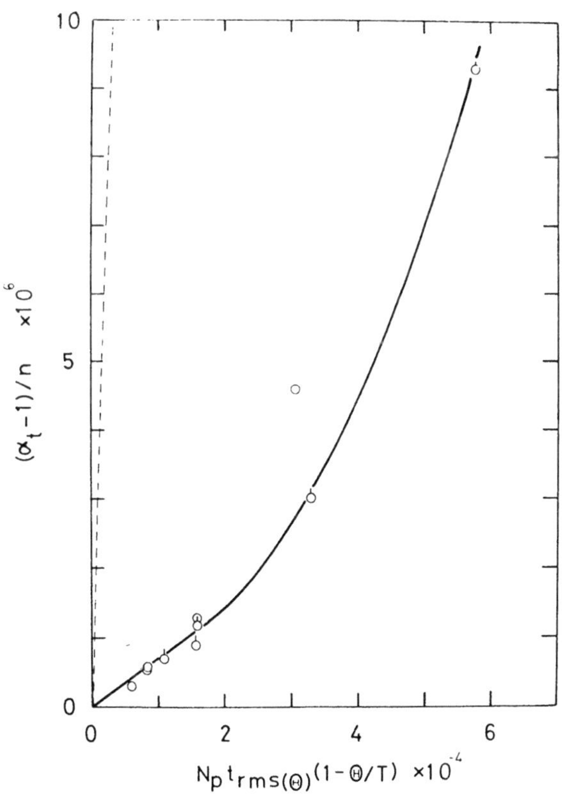

Fig. 18. Plot of $(\alpha_t - 1)/n$ vs. $N_p t_{rms(\Theta)} (1 - \Theta/T)$[74]. The dotted line is calculated from Eq. (D-6). Symbols are the same as in Fig. 17

Killmann and Wiegand[77] investigated the adsorption of poly(ethylene glycol) and poly(vinyl pyrrolidone) on a chromium plate from water and methanol solutions. The value of t_{rms} for poly(ethylene glycol) ($M_w = 40 \times 10^3$) in water came close to the root-mean-square end-to-end distance of the sample in bulk solution, whereas t_{rms} for poly(ethylene glycol) ($M_w = 6.13 \times 10^3$) in water and that for poly(vinyl pyrrolidone) ($M_w = 38 \times 10^3$) in water or methanol were much larger than those in bulk solutions. No comparison between theory and experiment was made.

D.1.2 Hydrodynamic Studies

Öhrn[26, 78, 79], among others, called attention to the anomalous viscosity behavior of polymer solutions at very low concentration. Plots of η_{sp}/C against C were found to curve either down or up at such concentrations, and the anomaly was attributed to the adsorption of polymer molecules onto the capillary wall. In order to calculate the thickness of the adsorbed polymer layer, Öhrn used the equation

$$\eta^*_{sp}/C = \eta_{sp}/C + 4L_H(\eta_{sp}/C + 1/C)(1/r) \quad \text{(D-7)}$$

where η^*_{sp}/C is the apparent reduced viscosity, η_{sp}/C the true reduced viscosity, L_H the hydrodynamic thickness of the adsorbed layer, and r the capillary radius. Equation (D-7) indicates that measurements at a fixed polymer concentration using viscometers of different capillary diameters allow us to estimate L_H. Öhrn[26] reported that L_H of polystyrene in toluene increased with rising C and reached a constant value. Tuijnman and Hermans[80] found an L_H as large as 5,000 Å for poly(vinyl acetate) ($M = 833 \times 10^3$) in toluene. This value is greater than the average diameter of the random coil in bulk solution.

Rowland and Eirich[81] estimated L_H from the rates of flow through a sintered glass disc being in adsorption equilibrium with the polymer solution, using the relation

$$\eta_{rel} J_s/J_0 = [(r - L_H)/\bar{r}]^4 \quad \text{(D-8)}$$

where η_{rel} is the relative viscosity, J_s and J_0 are the flow rates of polymer solution and solvent, respectively, and $\bar{r}$ is the average pore radius of the glass disc. The measured values of L_H are plotted as a function of the intrinsic viscosity $[\eta]$ for polystyrene, poly(methyl methacrylate), and poly(vinyl acetate) in Fig. 19. A linear relationship is observed between L_H and $[\eta]$, except for polystyrene in benzene, in which L_H tends to level off as $[\eta]$ is increased. They compared L_H with twice the radii of gyration of polymers in bulk solution and concluded that L_H of polystyrene in cyclohexane is approximately equal to the diameter of the polymer coil. On the other hand, the L_H for the other polymers were less that the coil dimensions in bulk solution. In addition to hydrodynamic thickness, Rowland and Eirich[81] determined for each of the above-mentioned systems the adsorbance onto the Pyrex No. 7740 glass powder and concluded from the calculations of the area occupied by spheres in the two-dimensional hexagonal closest packing that each of the polymer examined is adsorbed as a monolayer of somewhat compressed or interpenerated random coils.

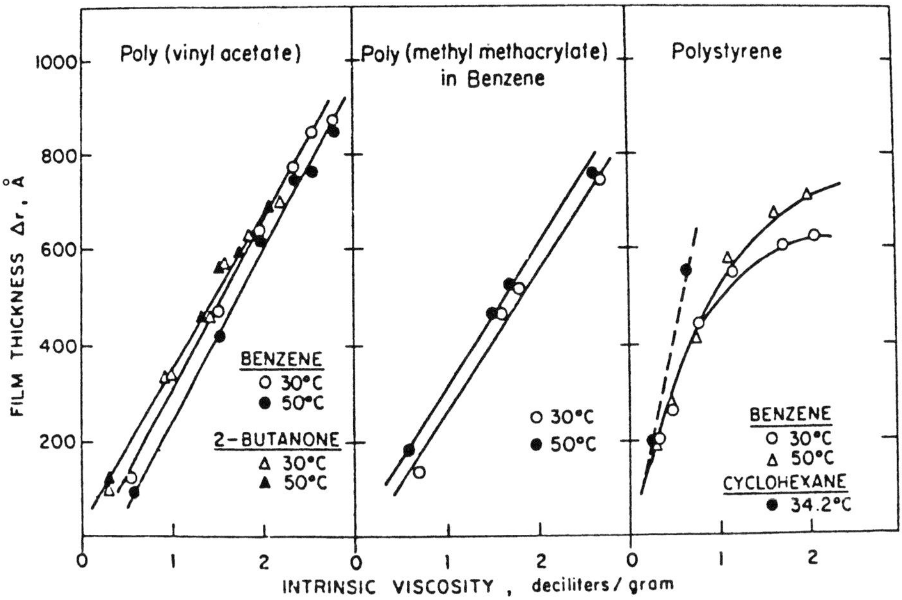

Fig. 19. Hydrodynamic thickness at high surface coverage as a function of intrinsic viscosity[81]

Recently, Priel and Silberberg[82] measured L_H for polystyrene in toluene and observed that the thickness changed considerably from a relatively small value to a very high plateau with increasing polymer concentration. The value in the plateau region increased with rising molecular weight and was close to the coil diameter calculated, using the Einstein equivalent sphere model.

Gramain[83] was the first to use a Millipore filter as an adsorbent. He measured L_H of adsorbed poly(vinyl acetate) as a function of molecular weight. L_H increased with time and reached a maximum after a period of 10–20 days. The equilibrium value of L_H was an increasing function of polymer concentration. At a fixed concentration this value varied linearly with the intrinsic viscosity of the polymer. Moreover, Gramain[83] investigated the influence of the pore size on L_H using filters with different pore diameters from 1,000 to 3,000 Å. For the same bulk concentration, L_H was twice as large for 3,000 Å pores as it was for 1,000 Å pores. The adsorbance was not measured. The values of L_H of adsorbed poly(vinyl acetate) on cellulose acetate were one half of those for the same polymer on Pyrex glass as investigated by Rowland and Eirich[8]. Gramain interpreted this difference as being due to a high polarity of cellulose acetate.

Very recently, Gramain and Myard[84] investigated the effect of the shear rate on L_H, using high molecular weight fractions of polyacrylamide and polystyrene. Polyacrylamide was hydrolyzed to various degrees. Five commercial filters of pore radii from 1.5 to 6 μm were used as porous media. Measurements were carried out on polyacrylamide in water and polystyrene in toluene.

Although the permeability reduction, which is defined as the ratio of flow rates before and after adsorption, was strongly affected by the nature of porous media, its dependence on the shear rate was similar for both polymers. A clear and reversible change of L_H

was observed when the shear rate was cyclically changed, indicating a reversible dilatation of the adsorbed layer. This dilatation increased with rising the molecular weight as well as the pore radius.

The measured values of L_H appreciably increased with increasing shear rate, and those at the highest shear rate were 3–6 times as large as the diameter of a free polymer in the bulk solution. To explain this result Gramain and Myard considered that there should exist a net circulation of solvent in the adsorbed layer to elongate adsorbed polymers axially. This effect reduces the number of trains to about 3 or 4 per polymer chain.

Fontana and Thomas[25)] estimated L_H from the sedimentation rate of carbon-black particles covered with poly(lauryl methacrylate) (PLMA) or stearyl methacrylate/N-vinyl-2-pyrrolidone copolymer (PAM/Å-VP). The values of L_H for PLMA were 20 to 40 Å, while those for PAM-VP were 210 ± 40 Å.

Garvey et al.[85)] made a similar sedimentation study on poly(vinyl alcohol) adsorbed on polystyrene latex particles. Adsorbance of the polymer was also measured. Both the thickness of the adsorbed layer and the adsorbance increased linearly with the square root of the molecular weight. The volume occupied by a polymer molecule in the adsorbed layer was approximately equal to that of the effective hydrodynamic sphere in bulk solution. However, the measured values of L_H were greater than the hydrodynamic diameters of the polymer coils in solution. Thus, it may be concluded that adsorbed poly(vinyl alcohol) assumes a conformation elongated in the direction normal to the surface.

Subsequently, Gavery et al.[86)] reinvestigated the adsorption of poly(vinyl alcohol) on polystyrene latex particles using three hydrodynamic methods: ultracentrifugation, photon correlation spectroscopy, and electrophoresis. The thickness obtained by photon correlation spectroscopy agreed with that obtained by ultracentrifugation. However, the thickness measured by electrophoresis was larger than those determined by the two other methods. They suggested that the assumptions made in the electrophoresis method be invalid, leading to the thickness dependent on electrolyte concentration. Photon correlation spectroscopy gave thicknesses increasing with the size of the polystyrene latex particles.

D.1.3 Other Approaches

Small-angle neutron scattering has proved to be a poweful tool for studying the semidilute regime where polymer coils overlap with one another. Extensive studies using this method are in progress. Its first application to adsorption studies was reported by Gebula et al.[87)] who investigated the density and size of laurate ions adsorbed on polystyrene latex particles. The technique has been extended to polymer adsorption by Barnett et al.[88)]. In the system they studied, the substrate was a partially deuterated monodisperse polystyrene latex which was prepared in an H_2O/D_2O dispersion medium from deuterated and protonated styrene monomers using a persulfate as the initiator. The adsorbed polymer was protonated poly(ethylene oxide) of molecular weight 50×10^3 prepared by anionic polymerization of ethylene oxide.

When a layer of thickness t is adsorbed on a particle of radius $r_0 (r_0 \gg t)$, it can be shown that the scattered intensity $I^0(q)$ is given by

$$I^0(q) = (\text{const}/q^2)\left[\int_0^t \varrho(z)\,e^{iqz}\,dz\right]^2 \tag{D-9}$$

where q is the scattering vector and $\varrho(z)$ the distribution function of adsorbed polymer segments. Barnett et al.[88] calculated the second moment σ_0 of the distribution from a plot of $\ln(q^2 I^0(q))$ against q^2, assuming a rectangular shape for $\varrho(z)$, i.e. a homogeneous adsorbed layer. The calculated σ_0 was approximately one half that estimated by photon correlation spectroscopy for different surface coverages of the polymer. This discrepancy remains to be elucidated.

Lyklema and Vliet[89] determined the equilibrium thickness t_0 of free liquid films stabilized by poly(vinyl alcohol) (PVA) adsorbed at the air-water interface. They estimated t_0 at different applied hydrostatic pressures by measuring the intensities of light reflected from the surface of the film to that of the silvery film. The t_0 values obtained increased with rising hydrostatic pressure and were extrapolated to zero pressure to obtain t_0 for a free film. The extrapolated t_0 should correspond to twice the thickness of the adsorbed PVA layer, but it far exceeded twice the latter determined by ellipsometry. The great difference was interpreted in terms of the presence of long dangling tails which are probably not to be seen by ellipsometry.

D.1.4 Attachment of Train Segments to Active Sites

To determine the conformation of adsorbed polymers the fraction of adsorbed polymer segments (p) and the fraction of the occupied surface sites (θ) are often measured. Fontana and Thomas[25] were the first to measure p and θ by IR spectroscopy. At present, the application of IR spectroscopy is limited to finely divided substrates, e.g. nonporous silica, and requires that the surface area and the number of surface sites (e.g. the silanol groups) per unit area are accurately known in advance. The adsorbed amount Γ of polymer per surface site can be determined from adsorbance A(g/cm^2) and the total area of the adsorbent. However, it can also be evaluated from the ratio θ/p.

Fontana and Thomas[25] determined p, θ, and Γ for poly (lauryl methacrylate) of molecular weights of 33×10^3 to 1190×10^3 adsorbed from n-dodecane and cis-decalin onto silica particles by observing IR shifts of the C = O bands in the polymer and the surface silanol group. They found that θ increased whereas p remained nearly constant with rising Γ.

Since Fontana and Thomas, IR spectroscopy has been extensively applied to the determination of p for various polymers adsorbed on the silica surface[90–99]. Recently, the EPR method has also been used to measure p[64, 100, 101]. Typical results from these studies are summarized in Table 1. As can be seen from this table, the dependence of p on adsorbance is classified into two categories: in one category, p decreases with increasing adsorbance and in the other, it is nearly independent of adsorbance. Qualitatively, the former indicates that the adsorbed polymer chain converts from a flattened to an extended conformation and the latter indicates that the adsorbed polymer undergoes no or only slight conformational changes.

Linden and Leemput[102] measured p, θ, and Γ for polystyrene adsorbed on the Aerosil 130 silica from cyclohexane and carbon tetrachloride solutions at 35 °C as a function of molecular weight. To minimize the complexity due to molecular weight

Table 1. Summary of published data

System	Temp. (°C)	M of polymer ($\times 10^{-3}$)	Adsorbance (mg/g)	p	Ref.
poly(lauryl methacrylate)/ n-dodecane/ silica	25.5	330	120 → 314	~0.35	25
poly(lauryl methacrylate)/ cis-decalin/silica	25.5	330	101 → 292	~0.37	25
poly(methyl methacrylate)/ chloroform/silica	25	230 2560	30 → 120 30 → 110	0.26 → 0.29 0.25 → 0.27	90
poly(4-vinyl pyridine)/chloro-form/silica	25	32 309 1690	60 → 200 100 → 200 70 → 200	0.55 → 0.17 0.34 → 0.29 0.45 → 0.19	90
polystyrene/ decalin/ silica	25	92	50 → 300	0.26 → 0.13	90
polystyrene/ trichloroethylene/ silica	25	92	60 → 100	0.26 → 0.02	90
poly(ethylene-o-tere-phthalate)/chloro-form/silica	35	25 54	60 → 70 25 → 100	~0.37 ~0.34	91
poly(methyl methacrylate)/ silica/ trichloroethylene	25	368	20 → 190	0.29 → 0.35	92
polystyrene/trichloro-ethylene/silica	25	246	40 → 110	0.28 → 0.13	92
poly(vinyl acetate)/ trichloroethylene/ silica	24	465	5 → 100	0.38 → 0.46	93
ethylene-vinyl acetate copolymer/ trichloroethylene/ silica	24	37.2	50 → 150	0.76 → 0.50	93
ethylene-vinyl acetate copolymer/ cyclohexane/silica	24	37.2	40 → 240	0.94 → 0.60	93
poly(oxyadipolyoxy-tetramethylene)/carbon tetrachloride/silica	29	4.4	20	1.0	94
poly(oxyadipoyloxy-hexamethylene)/carbon tetrachloride/silica	29	9.2	20	0.80	94

Table 1 (continued)

System	Temp. (°C)	M of polymer ($\times 10^{-3}$)	Adsorbance (mg/g)	p	Ref.
poly(oxylsuberoyloxy-hexamethylene)/carbon tetrachloride/silica	29	3.95	20	0.71	94
poly(oxysebacoyloxy-octamethylene)/carbon tetrachloride/silica	29	10 15	20 20	0.79 0.79	94 94
poly(oxycarbonyldeca-methylenecarbonyloxy-decamethylene)/carbon tetrachloride/silica	29	28	20	0.80	94
polystyrene/carbon tetrachloride/silica	29	260	20	0.34	94
poly(oxyadipoyloxy-hexamethylene/carbon tetrachloride/γ-Al_2O_3	30	90	37.5 → 179	0.34 → 0.082	95
poly(ethylene glycol)/carbon tetrachloride/γ-Al_2O_3	30	20	37.5 → 171	0.44 → 0.089	95
poly(oxydecamethyleneoxy-carbonyldecamethylene carbonyl)/carbon tetra-chloride/γ-Al_2O_3	30	2.5	11.1 → 65.8	0.76 → 0.56	95
poly(oxyethyleneoxy-succinyl)/chloroform/silica	25	0.69	20 → 90	0.43 → 0.35	96
poly(oxyhexamethylene-oxysuberoyl)/chloro-form/silica	25	3.2	20 → 90	0.67 → 0.48	96
poly(oxydecamethylene-oxydodecanediol)/chloroform/silica	25	5.84	20 → 70	0.81 → 0.64	96
poly(oxyethyleneoxy-1,2-phthalol)/chloroform/silica	25	2.58	20 → 90	0.38 → 0.35	96
poly(oxycarbonyloxy-1,4-phenylenepropane-2,2-diyl-1,4-phenylene)/chloroform/silica	25	27	20 → 90	0.56 → 0.31	96
poly(vinylpyrrolidone)/chloroform/silica	25 25	39.3 916	20 → 140 20 → 150	0.38 → 0.30 ~0.35	97
poly(vinylpyrrolidone)/D_2O/silica	25	39 916	20 → 110 20 → 130	~0.20 ~0.12	97
poly(vinyl acetate)/carbon tetrachloride/silica	25	25.5	300 → 928	0.4 → 0.2	98

Table 1 (continued)

System	Temp. (°C)	M of polymer ($\times 10^{-3}$)	Adsorbance (mg/g)	p	Ref.
poly(n-butyl methacrylate)/ carbon tetrachloride/ silica	25	3.2	10 → 107	0.8 → 0.75	98
	25	101	120 → 145	~0.8	
	25	4020	120 → 145	~0.8	
poly(caprolactone)/ carbon tetrachloride/ silica	25	33.0	110 → 211	0.75 → 0.30	98
polydimethylsiloxane/ carbon tetrachloride/ silica	29.5	2.08	10 → 33	0.93 → 0.47	99
		5.62	10 → 70	0.33 → 0.24	
		520	10 → 140	0.13 → 0.12	
vinylpyrrolidone-allylamine co-polymer/water/ silica	0–37	18	50 → 110	0.91 → 0.71	64,
		40	60 → 140	0.88 → 0.58	100
vinylpyrrolidone-allylamine co-polymer/chloroform silica	0–37	40	105 → 140	0.95 → 0.64	100
poly(vinylpyrrolidone)/ water/carbon	-10–40	40	75 → 200	0.95 → 0.67	101
poly(vinylpyrrolidone)/ aq.0.5M Na_2SO_3/ carbon	-10–40	40	150 → 380	0.97 → 0.90	101
poly(methyl metha-crylate)/chloroform/ silica	-10–40	85	15 → 40	1.0	101

distribution they used polymer samples with a narrow M distribution. At the theta point the adsorbance increased linearly with $M^{0.5}$ up to a molecular weight of 10×10^3 and then the exponent of M decreased monotonically to zero. The values of p obtained in the plateau region decreased with rising molecular weight whereas the values of θ increased and reached unity above molecular weights 2×10^3. It was found that the molecular weight dependence of θ, p and Γ was in better agreement with the theory of Silberberg[47] than that of Roe[49]. While the former theory is based on the loop-train model the latter assumes no particular conformation for the adsorbed polymer. The adsorbance obtained in carbon tetrachloride, which is a good solvent for polystyrene, was lower than that in cyclohexane. The molecular weight dependence of p in carbon tetrachloride showed an asymptotic behavior similar to that observed in cyclohexane. The values of p were slightly smaller than, but those of θ were just one half those in cyclohexane.

Very recently, Kawaguchi et al.[103] have studied the same system and compared the experimental data of p, θ, and Γ with the theory of Scheutjens and Fleer[48] and that of Silberberg[47]. The adsorption isotherms obtained were of the high affinity type. The molecular weight dependence of adsorbance in the plateau region displayed a behavior similar to that observed by Linden and Leemput. For each sample, p decreased with increasing adsorbance whereas θ exhibited the reverse trend. The value of p in the plateau region diminished with increasing molecular weight while the value of θ in the same region was at variance with the finding of Linden and Leemput, rising with molecular weight and reaching unity at the molecular weight of 110×10^3. Figures 20 and 21 illustrate that the experimental results are actually well fitted by either of the above-mentioned theories. Thus, Kawaguchi et al.[103] concluded that in order to obtain the

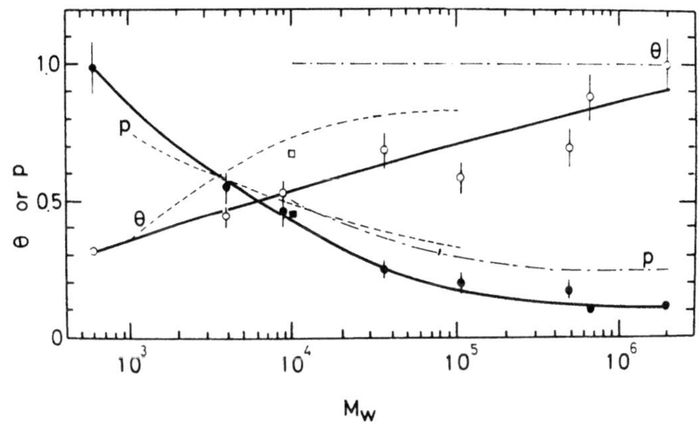

Fig. 20. Semilogarithmic plots of the fraction of occupied surface sites (○, θ) and the fraction of adsorbed segments (●, p) vs. molecular weight[103]. Dependence predicted at the theta temperature: (------) by the theory of Scheutjens and Fleer[48] for $\chi_s = 1.3$ and $\phi_* = 10^{-3}$; (—·—·—·—) by the theory of Silberberg[47] for $\gamma_s\gamma_B = 0.1$, $\chi_s = \infty$, and $\phi^* = 10^{-3}$; (□, θ; ■, p) by the theory of Silberberg[47] for $\gamma_s\gamma_B = 0.1$, $\chi_s = 1.3$ and $\phi^* = 10^{-3}$

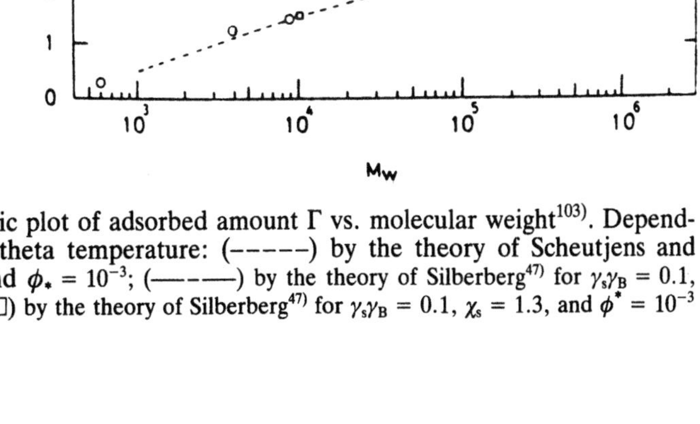

Fig. 21. Semilogarithmic plot of adsorbed amount Γ vs. molecular weight[103]. Dependence predicted at the theta temperature: (-----) by the theory of Scheutjens and Fleer[48] for $\chi_s = 1.3$ and $\phi_* = 10^{-3}$; (—·—·—) by the theory of Silberberg[47] for $\gamma_s\gamma_B = 0.1$, $\chi_s = \infty$, and $\phi^* = 10^{-3}$; (□) by the theory of Silberberg[47] for $\gamma_s\gamma_B = 0.1$, $\chi_s = 1.3$, and $\phi^* = 10^{-3}$

predominant conformation of adsorbed polymers, i.e. whether it is a loop-train or a loop-train-tail, not only data of p, θ, and Γ but also those of the thickness of the adsorbed layer must be determined experimentally.

D.1.5 Heats of Adsorption

Killmann and Eckart[65] were the first to determine adsorption enthalpies by calorimetry. They measured the enthalpies for poly(ethylene glycol) adsorbed from carbon tetrachloride, benzene, water, and methanol solutions onto finely grained silica. The obtained values increased linearly with adsorbance in the order methanol, water, benzene, and carbon tetrachloride.

Integral adsorption enthalpies ΔH for the adsorption of poly(vinyl acetate), poly(n-butyl methacrylate), and polycaprolactone from carbon tetrachloride solutions on Aerosil 200 silica were measured by Korn and Killmann[98], who used IR spectroscopy for the determination of p. The values obtained for the three polymers exhibited the same characteristic dependence on adsorbance as that observed for poly(ethylene glycol). If the enthalpy of polymer-solvent and polymer-polymer interactions is neglected, the net binding adsorption enthalpy $\Delta H_{p,A}$ can be calculated from

$$\Delta H_{p,A} = \Delta H/(Ap) + \Delta H_{L,A} \tag{D-10}$$

where $\Delta H_{L,A}$ is the enthalpy associated with the interaction between solvent and adsorbent. The calculated values of $\Delta H_{p,A}$ were independent of A and M.

Hair[104] related $\Delta H_{p,A}$ to the frequency shift $\Delta\nu$ of a specific group of the adsorbent by $\Delta H_{p,A} \propto (\Delta\nu)^{1/2}$, and Korn et al.[105] examined this relationship. Although the expected linear relation was found for low molar mass monofunctional adsorbates, $\Delta H_{p,A}$ for the investigated polymers decreased considerably below the straight line for this relation. Korn et al. attributed this discrepancy to three facts. First, more than one adsorbed solvent molecule were displaced when a polymer segment was adsorbed on the silica surface. Second, since the microcalorimetric system used was designed to detect relatively fast thermal effects (completed within several minutes) but polymer adsorption was a very slow process, the heat evolved by unfolding or rearrangement of the polymer coming into contact with the surface was not detected. Third, no attention was paid to interactions between adsorbed polymers.

D.2 Adsorption of Branched Polymers

Since a branched polymer has more than two chain ends, the determination of both the thickness of the adsorbed polymer and the adsorbance is of particular interest. Kawaguchi and Takahashi[106] investigated well-characterized comb-branched polystyrene adsorbed on a chromium plate from cyclohexane solution at the theta point by ellipsometry and compared the results with those for the corresponding linear polystyrene.

The adsorption isotherms obtained as well as the plot of the thickness of the adsorbed layer against the bulk polymer concentration were similar to those for linear polystyrene, but the adsorbance in the plateau region for the branched polymer was larger than that for the linear one when compared at the same molecular weight. The root-mean-square thickness $t_{b,\,rms}$ of branched polystyrene was smaller than that for the linear one and only slightly larger than the diameter of the unperturbed branched polystyrene in solution. This fact suggests that comb-branched polystyrene is adsorbed as a slightly distorted random coil. Moreover, $t_{b,\,rms}$ is close to $t_{\ell,\,rms}g_s^{1/2}$, as shown in Fig. 22, where $t_{\ell,\,rms}$ is the root-mean-square thickness of linear polystyrene of the same molecular weight and $g_s^{1/2}$ the ratio of the root-mean-square radii of gyration of branched and linear polymers with the same molecular weight[107]. The ratio of adsorbances A_b/A_l of branched and linear polystyrenes was approximately equal to g_s^{-1}. Summing up, comb-branched polystyrene is adsorbed as a slightly distorted random coil with a thickness and an adsorbance governed primarily by the g_s factor.

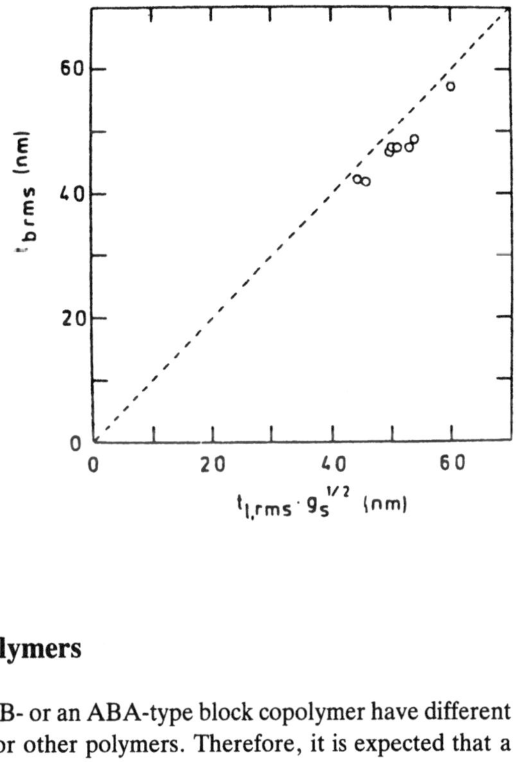

Fig. 22. Plots of $t_{b,\,rms}$ against $t_{\ell,\,rms}$ $g_s^{1/2}$ [106]

D.3 Adsorption of Block Copolymers

In general, A and B subchains in an AB- or an ABA-type block copolymer have different solubilities or affinities for a solvent or other polymers. Therefore, it is expected that a block copolymer is surface-active when dissolved in a suitable solvent or mixed in polymer melts[108]. This property of block copolymers is now utilized to stabilize or flocculate colloidal dispersions. Blocks A, which are insoluble in a given solvent, are anchored in an insoluble polymer particle, and blocks B, which are soluble in the solvent, form a surface layer around the particle.

Dawkins and Taylor[109] dispersed poly(methyl methacrylate) (PMMA) or polystyrene (PS) particles in n-alkanes stabilized by AB block copolymers of styrene and dimethylsiloxane. In these cases, styrene blocks act as anchors and dimethylsiloxane blocks give a surface layer. The thickness δ of the dimethylsiloxane layer was determined by viscosity measurements as a function of the molecular weight of dimethylsiloxane blocks.

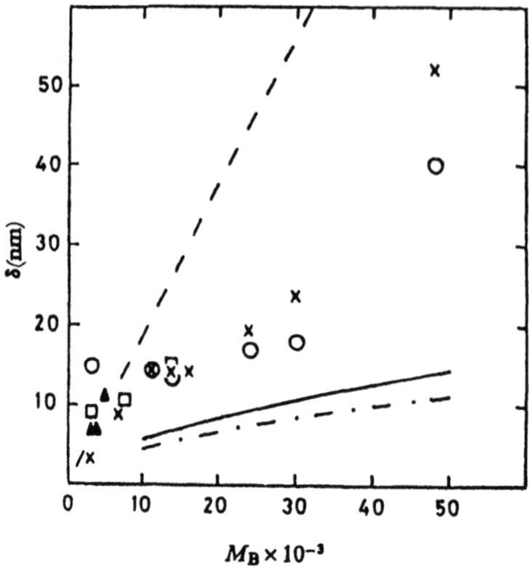

Fig. 23. Variation of surface layer thickness with molecular weight of the stabilizing polydimethylsiloxane (PDMS) chain. Hydrodynamic thickness δ[109]: PMMA particles (○), PS particles (□), micellar dispersions (Δ) from viscosity data; ×, thickness h from surface coverage data of PMMA particles assuming a prolate ellipsoid model for the PDMS chain; (-----) extended chain model; (——) twice the root-mean-square radius of gyration for the random-coil model of the PDMS chain; (— — —) hydrodynamic diameter for the random coil-model of the PDMS chain

In Fig. 23, δ is plotted against M_B, the molecular weight of dimethylsiloxane blocks, for various dispersions. As can be seen, above $M_B = 10 \times 10^3$ δ falls between the two limiting lines corresponding to the fully stretched chain model and the random coil model, while below $M_B = 10 \times 10^3$, δ is closer to the former model than to the latter.

From the adsorbance of dimethylsiloxane blocks, the mean separation y between adjacent dimethylsiloxane chains was calculated and found to be almost equal to the radius of gyration $\langle s^2 \rangle^{1/2}$ of the dimethylsiloxane chain calculated from intrinsic viscosity.

Since n-alkanes are good solvents for poly(dimethylsiloxane), the adjacent dimethylsiloxane chains may not interpenetrate, as a result of excluded volume effect. Thus, the shape of a dimethylsiloxane chain may be represented by a prolate ellipsoid with y as the minor axis. The major axis h may be calculated by equating the volume of ellipsoid to $(y/2)^2(h/2)$ to $\langle s^2 \rangle^{3/2}$. The h value so calculated agreed reasonably well with the value of δ. Therefore, the shape of the stabilizing dimethylsiloxane chain should be close to an ellipsoid.

D.4 Adsorption of Polyelectrolytes

Adsorption of polyelectrolytes onto solid surfaces are not yet explored as extensively as that of non-ionic polymers, and most studies are limited to adsorbance measurements.

The adsorption of sodium polyacrylate on kaolinite was investigated by Michaels and Morelos[110] in the pH range 5–8. Adsorbance was found to decrease with rising pH and to increase in proportion to $M^{0.05}$. Schmidt and Eirich[111] dealt with a series of vinyl acetate-crotonic acid copolymers adsorbed on anatase as a function of the charge density of copolymer and the pH of solution. Adsorbance largely increased as the crotonic acid content was raised and also as the pH was lowered.

The adsorption of ionized poly(4-vinylpyridine) onto a glass surface was investigated by Peyser and Ullman[112], who found that adsorbance decreased with increasing ionization and increased with rising concentration of added salt. These results are in qualitative agreement with theoretical predictions[23, 57].

Recently, Bartels and Arends[113] studied the adsorption of poly(4-vinylpyridinium fluoride) with different hexadecyl group content on hydroxyapatite. Adsorbance decreased as the hexadecyl content, i.e. the charge density, was increased. Desorption experiments showed that the adsorption of this polyelectrolyte in water is essentially irreversible. However, the polymer partially desorbed when excess calcium ions were added. Bartels and Arends concluded that adsorption of poly(4-vinylpyridinum fluoride) occurs as a result of the uptake of fluoride ions by hydroxyapatite which releases phosphate ions into water. They also suggested that this adsorption phenomenon can be interpreted in terms of an ion-exchange mechanism.

The first ellipsometric measurement of the thickness of the adsorbed layer and the adsorbance of a polyelectrolyte and a negative adsorbance of salt onto a solid surface was reported by Takahashi et al.[114] They measured the adsorption of sodium poly(acrylate) ($M = 950 \times 10^3$) onto a platinum plate as a function of the concentration of added sodium bromide. In an aqueous polyelectrolyte solution with an added simple salt, the bulk phase is a three-component system which consists of a polyelectrolyte, a simple salt, and water. The adsorbed layer on the solid surface is a three-component phase as well. The adsorbance of polyelectrolytes thus cannot easily be determined from measurements of the refractive index n_f of the adsorbed phase. Hence, it was assumed that the adsorbed layer is a homogeneous layer of thickness t and further that n_f is represented by the Lorenz-Lorentz equation as follows:

$$n_f = [(\overline{M} + 2d\overline{R})/(\overline{M} - d\overline{R})]^{1/2} \tag{D-11}$$

or

$$\overline{R} = (n_f^2 - 1)\overline{M}/(n_f^2 + 2)d \tag{D-12}$$

where $\overline{M}$ is the mean molecular weight, $\overline{R}$ the mean molar refractivity, and d the density of the adsorbed layer. $\overline{M}$ and $\overline{R}$ are given by

$$\overline{M} = \sum_i (X_i M_i) \tag{D-13}$$

$$\overline{R} = \sum_i (X_i R_i) \tag{D-14}$$

where X_i denotes the mole fraction of component i. The density d is expressed by

$$d = d_0 + (M_+ - d_0 V_+^0)\, C_+/1000 + (M_- - d_0 V_-^0)\, C_-/1000 + (M_p - d_0 V_p^0)\, C_p/1000 \tag{D-15}$$

where d_0 is the density of water, V_+^0, V_-^0 and V_p^0 are the apparent molar volumes, M_+, M_- and M_p the molecular weights, and C_+, C_- and C_p the concentrations of cation, anion and polyion, respectively. The Donnan equilibrium should be established between the bulk phase and the adsorbed layer, and this condition gives

$$C_s^0(C_s^0 + \nu\phi_p C_p^0) = [C_s^0 + (\Gamma_s + \nu\phi_\sigma \Gamma_p/t)](C_s^0 + \Gamma_s/t) \tag{D-16}$$

where C_s^0 and C_p^0 are the molar concentrations of uni-univalent salt and polyelectrolyte in the bulk phase, respectively, ν is the number of charges per polyion, ϕ_p and ϕ_σ are the osmotic coefficients for the salt-free bulk polyelectrolyte phase and the adsorbed layer, and Γ_p and Γ_s are the adsorbances of polyelectrolyte and simple salt in mol/cm^2. Eq. (D-16) was first derived by Frommer and Miller[115] assuming the additivity rule for the osmotic factors[116, 117].

From the refractive indices and densities for sodium polyacrylate in aqueous NaBr the molar refractivities and the apparent molar ion volumes were calculated. With $\phi_p = \phi_\sigma = 0.2$[118] and the measured values of n_f, t, C_p^0, and C_s^0, Γ_p and Γ_s were evaluated by solving Eqs. (D-11), (D-13), (D-14), (D-15), and (D-16).

The adsorption isothems obtained first rise with the polyelectrolyte concentration and level off to a plateau, as shown in Fig. 24. The adsorbance at the polymer concentration of 0.1 g/dl, which is well in the plateau region, decreases as the salt concentration is lowered, and it varies linearly with the square root of the salt concentration, as shown in Fig. 25. This linear relationship agrees with the theoretical prediction by Hesselink[23].

The adsorbance of NaBr, Γ_s, was large and negative. The Donnan exclusion factor $\Gamma_{ex} = \Gamma_s/\Gamma_p$ was in fair agreement with that calculated from the wire model of Devore and Manning[119] for the polyelectrolyte in salt solution.

The thickness of the adsorbed polymer layer increases gradually with the sodium polyacrylate concentration and is constant above 0.03 g/dl, as is illustrated in Fig. 26. The thickness in the plateau region (0.1 g/dl) rises with decreasing salt concentration. The low adsorbance and the high thickness at low salt concentration are due to the electrostatic repulsion between charged groups of polyions. As the ionic strength is increased, both the intra- and interpolyion interactions diminish so that a larger adsorbance and a smaller thickness should be obtained.

To determine the conformation of adsorbed sodium polyacrylate, we first assume a loop-train conformation. Then, t_{rms} at the theta point can be evaluated by Eq. (B-108). Since 1.5N NaBr at 15 °C is the theta solvent for sodium polyacrylate[120], this calculation

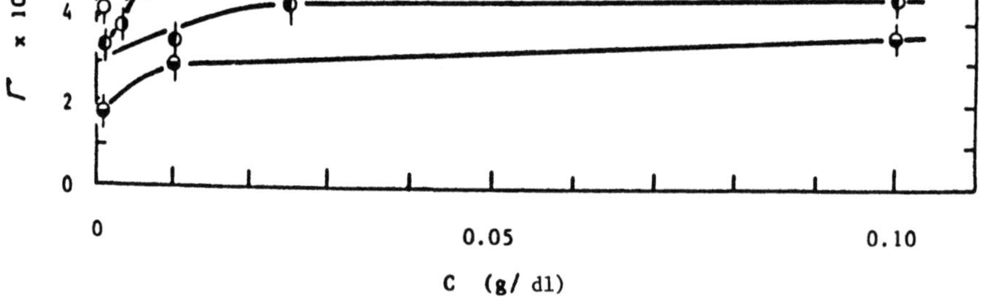

Fig. 24. Adsorption isotherms of sodium polyacrylate in NaBr solutions[114]; salt concentration, C_s^0: (○) 1.5 N; (◑) 0.5 N; (◐) 0.1 N; (◒) 0.005 N

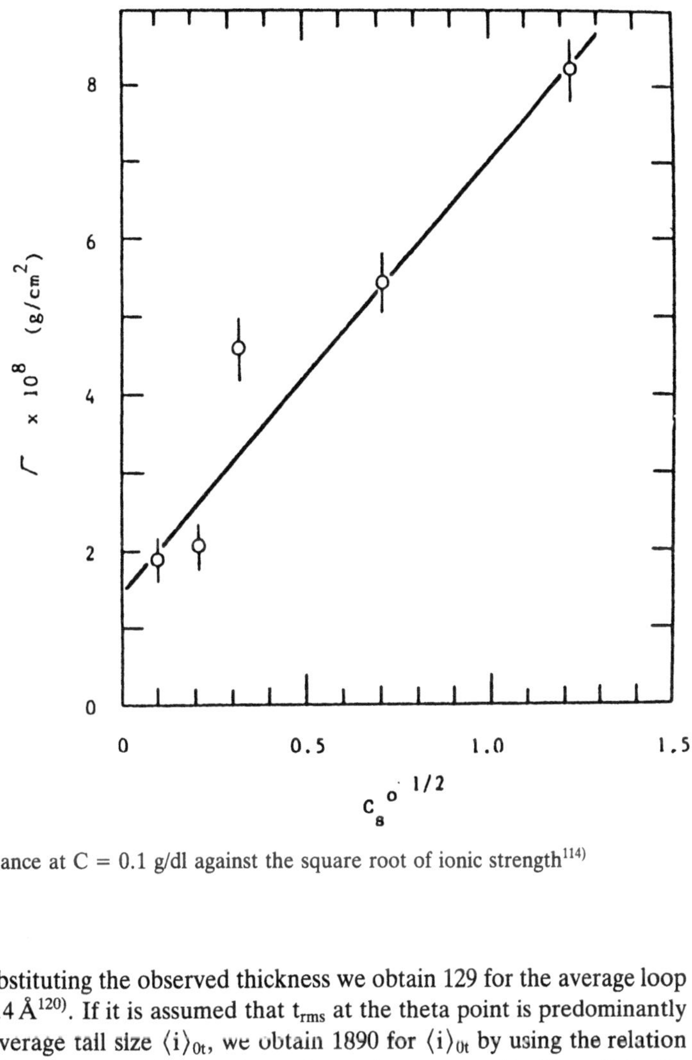

Fig. 25. Plot of adsorbance at C = 0.1 g/dl against the square root of ionic strength[114)]

is meaningful. By substituting the observed thickness we obtain 129 for the average loop size $\langle i \rangle_{0\ell}$ and b = 7.4 Å[120)]. If it is assumed that t_{rms} at the theta point is predominantly determined by the average tail size $\langle i \rangle_{0t}$, we obtain 1890 for $\langle i \rangle_{0t}$ by using the relation $t_{rms} = (7/6 \langle i \rangle_{0t})^{1/2} b$. For the one tail-one train conformation, assumed by Hesselink, the tail fraction f^t in the plateau region of the adsorption isotherm is given by Eq. (B-156). Since $\sum_{i=1}^{n} i^{1/2} \cong (2/3) n^{3/2}$ and $\sum_{i=1}^{n} i^{-1/2} \cong 2 n^{1/2}$, we then get 1/3 for f^t. The corresponding tails size is 3,300 since the degree of polymerization of the polyelectrolyte examined is 10×10^3. This value is about twice the tail size $\langle i \rangle_{0t} = 1890$. Hoping to find which of the loop-train and loop-train-tail conformations is more relevant, Takahashi et al.[114)] compared the expansion factor α_t of the adsorbed layer with the excluded volume function of the polyelectrolyte in salt-added solution.

Empirically, the expansion factor α_s for the radius of gyration of a polyion, due to electrostatic volume exclusion, is given by[120–123)]

$$\alpha_s^3 - 1 = k_\eta (M/C_s^0)^{1/2} \qquad \text{(D-17)}$$

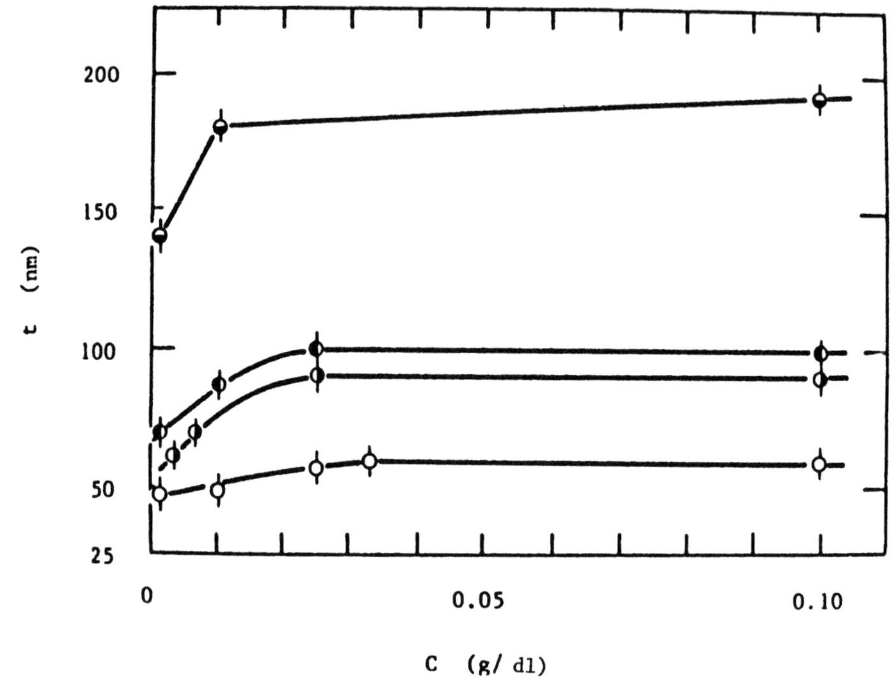

Fig. 26. Plots of thickness of the adsorbed layer vs. polyelectrolyte concentration at various ionic strengths[114]. Symbols are the same as in Fig. 24

where M is the molecular weight of the polyelectrolyte, C_s^0 the concentration of the added salt, and k_η a constant determined from intrinsic viscosity measurements. By assuming that the same form as Eq. (D-17) holds for the adsorbed polymer layer, the values of $\alpha_s^3 - 1$ for $\langle i \rangle_{0t} = 1890$ and $\langle i \rangle_{0\ell} = 129$ are computed and compared with observed $(\alpha_t^3 - 1)$ in Fig. 27. The agreement is good for the average tail size but very poor for the average loop size. This fact indicates that the tail portion of the adsorbed polyelectrolyte governs the adsorbed thickness. However, except for dangling tails, the polyelectrolyte chain perhaps assumes a loop-train conformation.

Pefferkorn et al.[124] measured the hydrodynamic thickness of an alternating copolymer of maleic acid and ethyl vinyl ether adsorbed on the pore walls in cellulose ester filters as a function of the molecular weight and the ionic strength of added NaCl. The measured thicknesses of the adsorbed polyelectrolyte were larger for higher molecular weights and lower ionic strengths while the adsorbances were smaller for lower ionic strengths. The thickness at the theta point varied linearly with intrinsic viscosity. Assuming that the segment distribution in the adsorbed polymer layer is exponential[59], Pefferkorn et al. derived between α_t and the excluded volume ν the following relation:

$$\alpha_t^3 - \alpha_t = n_{0\theta}^2 b_\theta \nu / 12m \qquad \text{(D-18)}$$

where $n_{0\theta}$ is the segment density on the pore walls under the theta condition, b_θ the average loop length under the same condition, and m the number of loops per cm^2 of the

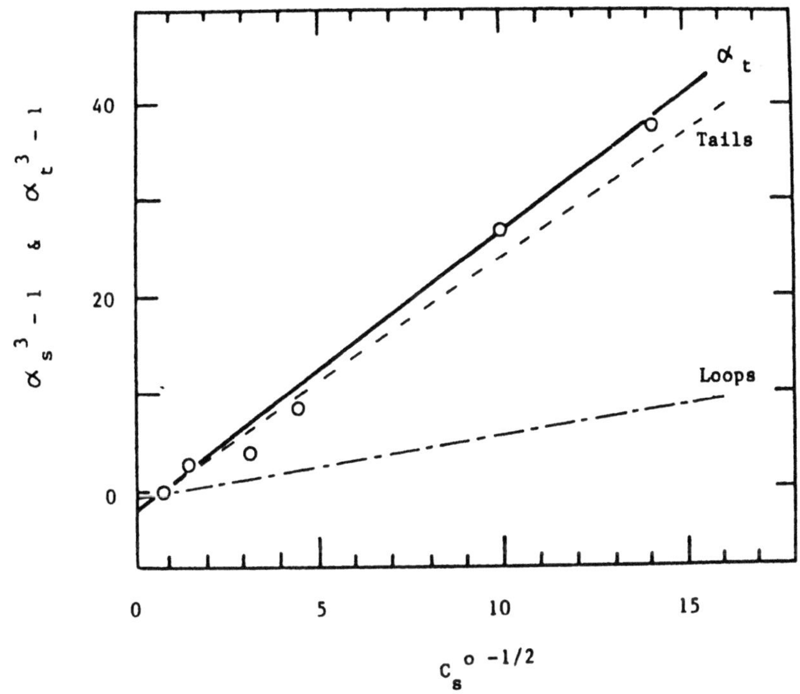

Fig. 27. Comparison of plots for $\alpha_t^3 - 1$ and $\alpha_s^3 - 1$ vs. the inverse square root of ionic strength[114]: (-------) $\alpha_s^3 - 1$ for the tail size $\langle i\rangle_{0t}$ = 1,890; (———) $\alpha_s^3 - 1$ for the loop size $\langle i\rangle_{0\ell}$ = 129

wall. They found that $b_\theta \propto M^{0.5}$ whereas $\alpha_t^3 - \alpha_t$ was proportional to M. These results led them to the conclusion that $n_{0\theta}^2/m$ should vary in proportion to the square root of M.

D.5 Adsorption of Block Polyelectrolytes

The adsorption behavior of AB- or ABA-type block copolymers in which block A is polyelectrolytic and block B hydrophobic is very interesting. As expected, these polymers serve as dispersants, micelle-forming agents and surface-active agents.

Takahashi et al.[67] prepared ionene-tetrahydrofuran-ionene (ITI) triblock copolymers and investigated their surface activities. Surface tension-concentration curves for salt-free aqueous solutions of ITI showed that the critical micelle concentration (CMC) decreased with increasing mole fraction of tetrahydrofuran units in the copolymer. This behavior is due to an increase in hydrophobicity. The adsorbance and the thickness of the adsorbed layer for various ITI at the air-water interface were measured by ellipsometry. The adsorbance was also estimated from the Gibbs adsorption equation extended to aqueous polyelectrolyte solutions. The measured and calculated adsorbances were of the same order of magnitude. The thickness of the adsorbed layer was almost equal to the contour length of the ionene blocks. The intramolecular electrostatic repulsion between charged groups in the ionene blocks is probably responsible for the full extension of the polymer chain. The conclusion was that the adsorbed ITI block copolymer assumes a hairpin conformation.

Kawaguchi et al.[125] prepared an ionene-oxyethylene-ionene (IEI) triblock copolymer with the molecular weight 72×10^3 and measured its surface tension in aqueous KBr. They also determined by ellipsometry the adsorbance and the thickness of the adsorbed polyelectrolyte layer at the air-KBr solution interface as a function of the KBr concentration. The data obtained indicate that this copolymer is surface-active and that the effect of added KBr on the surface tension is stronger than in the case of polyoxyethylene (POE).

The occupied areas S_L per adsorbed polymer at the air-water interface estimated from adsorbances for IEI and POE were almost equal and independent of the KBr concentration. The thickness of the adsorbed POE did not exceed 15 Å while that of the adsorbed IEI increased with decreasing KBr concentration and was about 1.5 times as large as the root mean-square end-to-end distance of the ionene homopolymer with the same M in bulk solution. It was concluded that POE, which is non-ionic and hydrophilic, is adsorbed in a nearly flattened conformation with short loops and trains. On the other hand, IEI is adsorbed a in tail-train-tail conformation in which the oxyethylene block lies flat on the air-water interface and the ionene tails are elongated to the bulk solution.

D.6 Adsorption of Polyampholytes

Maternaghan and Ottewill[126] studied by ellipsometry the adsorption of gelatin on the (111) face of a single silver bromide crystal at 45 °C. The gelatin used was of photographic grade. The adsorption from solutions containing 10^{-2} mol/l potassium nitrate was investigated in the pH range 3.5–8.5.

The adsorbance and the thickness of the adsorbed layer increased to a maximum at a pH close to the isoelectric point of the gelatin (pH = 4.8). On the acid side of the isoelectric point, these quantities decreased sharply with diminishing pH. On the alkali side they decreased between pH = 5 and 8.5 and were constant above pH = 8.5.

Eirich and Kudish[37, 127] studied the adsorption of two gelatins on glass at 30 °C. One sample was a single-stranded alkali-treated calfskin gelatin with $M = 90 \times 10^3$, and the other was an acid-treated pigskin „parent“ gelatin with $M = 300 \times 10^3$ in which three strands of the original collagen entangle together. The adsorption was studied in the pH range 2.5–11. The hydrodynamic thicknesses of the adsorbed layer determined by the capillary method were compared favorably with the radii of gyration calculated from instrinsic viscosities. They exhibited a maximum at a pH close to the isoelectric point. At pH below and above the isoelectric point, the thicknesses of the adsorbed layer for both gelatins were smaller than those at the isoelectric point. The pH effects on the thickness were similar to that observed by Maternaghan and Ottewill[126] using ellipsometry.

D.7 Adsorption of Rod-Like Macromolecules

Polypeptides undergo a conformational change from helix to random coil when the solvent composition, pH or temperature is varied. Adsorption of polypeptides such as poly(β-benzyl-L-glutamate), poly-proline, and polyhydroxyproline on glass was studied by Eirich and Chao[37, 128] using the hydrodynamic method. Adsorbances were also measured. For a polypeptide existing in a helical conformation, which may be regarded as a stiff rod, both adsorbance and thickness measurements revealed either a very thin or a very thick adsorbed layer. Moreover, the area occupied by one polypeptide chain in the very thick adsorbed layer was almost identical to that of its monomer unit. Therefore, it can be concluded that a helical polypeptide is adsorbed either by attaching its one end to or placing its entire length on the adsorbing surface, as schematically depicted in Figs. 1-f and 1-g. The decrease of the hydrodynamic thickness with diminishing helicity was also reported.

E. Conclusions

This review has surveyed the present status of theoretical and experimental studies on the adsorption of macromolecules, with particular reference to the determination of conformations of adsorbed macromolecules. No definitive experimental method is as yet available for this determination. At present, the conformation of adsorbed macromolecules can only be inferred from a comparison of experimental data of such properties as the adsorbed amount Γ, the fraction p of adsorbed trains and the thickness of the adsorbed layer with the existing theories formulated on various conformational models as depicted in Fig. 1.

From studies performed with well characterized substrates and polystyrene with a narrow M distribution, the measured values of Γ, p, and θ at the theta point have been found to agree closely with the theories of Silberberg and Scheutjens and Fleer. Furthermore, it has been shown that the measured root-mean-square thickness of the adsorbed layer can be predicted semiquantitatively by the loop-train-tail conformation model.

The theoretical description of excluded volume effects on the adsorption from good solvents is still unsatisfactory. The scaling theory for polymer adsorption has not yet been subject to experimental tests.

For the adsorption of polyelectrolytes, the existing theories predict only qualitatively the observed behavior, and more theoretical improvements remain to be made.

Measurements of hydrodynamic thickness L_H have been performed by many investigators and, in most cases, the measured L_H were almost twice the radii of gyration of polymer coils in bulk solution. It is desirable to clarify the theoretical relationship between L_H and the root-mean-square thickness of the adsorbed polymer layer. Some progress in this direction has been made recently.

For block copolymers in which one block acts as anchor and the other block dangles into bulk solution, the conformation of the adsorbed polymer was readily determined due to the characteristic molecular structure. It was shown that the conformation is either tail-train or tail-train-tail, depending on whether the block copolymer is AB or ABA, and that the tails are more elongated than the end-to-end distance of the block chain in bulk solution.

The present methods inferring the polymer conformation from data on Γ, p, θ, and t_{rms} must be refined further. For example, spectroscopic ellipsometry which should allow Γ, p, θ, and t_{rms} to be determined simultaneously might be a powerful tool.

Many studies have revealed that the adsorption behavior of macromolecules is very specific for individual polymer-solvent-substrate systems. Hence, the formulation of a general theory of polymer adsorption might be extremely difficult. Nevertheless, we believe that more experimental studies with well characterized polymers and surfaces are imperative of this difficulty to be overcome.

Acknowledgments. The authors are grateful for the help of Prof. Tadaya Kato who read and criticized various sections of this review. The technical assistance of Messrs. Hideyuki Hirota and Kazuhisa Hayakawa in performing ellipsometry is deeply appreciated. Professor Fujita's invitation to prepare this review and his linguistic comments are acknowledged with thanks.

F. References

1. Frisch, H, L., Simha, R., Eirich F. R.: J. Chem. Phys. *21*, 365 (1953)
2. Simha, R., Frisch, H. L., Eirich, F. R.: J. Phys. Chem. *57*, 584 (1953)
3. Frisch, H. L., Simha, R.: J. Phys. Chem. *58*, 507 (1954)
4. Frisch, H. L.: J. Phys. Chem. *59*, 633 (1955)
5. Frisch, H. L., Simha, R.: J. Phys. Chem. *27*, 702 (1957)
6. Simha, R.: J. Polym. Sci. *29*, 3 (1958)
7. Silberberg, A.: J. Phys. Chem. *66*, 1872 and 1884 (1962); J. Chem. Phys. *46*, 1105 (1967)
8. DiMarzio, E. A.: J. Chem. Phys. *42*, 2101 (1965)
9. DiMarzio, E. A., McCrackin, F. L.: J. Chem. Phys. *43*, 539 (1965)
10. Roe, R-J.: Proc. Natl. Acad. Sci. *53*, 50 (1965)
11. Roe, R-J.: Chem. Phys. *43*, 1591 (1965)
12. Roe, R-J.: J. Chem. Phys. *44*, 4264 (1966)
13. Motomura, K., Matuura, R.: J. Chem. Phys. *50*, 1281 (1969)
14. Chan, D. et al.: J. Chem. Soc. Faraday Trans. II, *71*, 235 (1975)
15. Rubin, R. J.: J. Chem. Phys. *43*, 2392 (1965)
16. Rubin, R. J.: J. Res. Natl. Bur. Std. *69 B*, 301 (1965)
17. Clayfield, E. J., Lumb, E. C.: J. Colloid Interface Sci. *22*, 269 and 285 (1966)
18. McCrackin, F. L.: J. Chem. Phys. *47*, 1980 (1967)
19. Lax, M.: Macromolecules *7*, 660 (1974)
20. Lal, M. et al.: Configurational behavior of isolated chain molecules adsorbed from athermal solutions, in: Adsorption at Interfaces (ed.) Mittal, K. L., p. 16, ACS Symp. Ser. 8, Washington D. C., Amer. Chem. Soc. 1975
21. Feigin, R. I., Napper, D. H.: J. Colloid Interface Sci. *71*, 117 (1979)
22. Kleinschmidt, A., in: Methods in Enzymology (eds.) Grossman, L., Moldave, K., Vol. 12 B, p. 361, New York, N. Y., Academic Press 1968
23. Hesselink, F. Th.: J. Electroanal. Chem. *37*, 317 (1972)
24. De Gennes, P. G.: J. Phys. *37*, 1445 (1976)
25. Fontana, B. J., Thomas, J. R.: J. Phys. Chem. *65*, 480 (1961)
26. Öhrn, O. E.: J. Polym. Sci. *17*, 137 (1955)
27. Stromberg, R. R., Passaglia, E., Tutas, D. J.: J. Res. Natl. Bur. Std. *67A*, 431 (1963)
28. Hughes, R. E., von Frankenberg, C. A.: Ann. Rev. Phys. Chem. *14*, 291 (1963)
29. Pataţ, F., Killmann, E., Schliebener, C.: Adv. Polym. Sci. *3*, 332 (1964)
30. Kipling, J. J.: Adsorption from Solutions of Non-Electrolytes, p. 134, New York, N. Y., Academic Press 1965
31. Stromberg, R. R.: Adsorption of polymers, in: Treatise on Adhesion and Adhesives (ed.) Patrick, R. I., Vol. 1, p. 69, New York, N. Y., Marcel Dekker 1967
32. Silberberg, A.: Pure Appl. Chem. *26*, 583 (1971)

33. Fontana, B. J.: Adsorption of biological analog molecules on nonbiological surfaces. Polymers, in: The Chemistry of Biosurfaces, (ed.) Hair, M. L., Vol. 1, p. 83, New York, N. Y., Marcel Dekker 1971
34. Ash, S. G.: Polymer adsorption at the solid liquid interface, in: Colloid Science, Vol. 1, p. 103, London, The Chemical Soc. 1973
35. Lipatov, Yu, S., Sergeeva, L. M.: Adsorption of Polymers, New York, N. Y., Wiley & Sons 1974
36. Vincent, B.: Advan. Colloid Interface Sci. *4*, 193 (1974)
37. Eirich, F. R.: J. Colloid Interface Sci. *58*, 423 (1977)
38. Chandrasekhar, S.: Rev. Mod. Phys. *15*, 1 (1943)
39. Hesselink, F. Th.: J. Phys. Chem. *73*, 3488 (1969)
40. Hesselink, F. Th.: J. Phys. Chem. *75*, 65 (1971)
41. Hoeve, C. A. J., DiMarzio, E. A., Peyser, P.: J. Chem. Phys. *42*, 2558 (1965)
42. Hoeve, C. A. J.: J. Chem. Phys. *44*, 1505 (1966)
43. Hoeve, C. A. J.: J. Chem. Phys. *43*, 3007 (1965)
44. Hoeve, C. A. J.: J. Polym. Sci. Part C., *30*, 361 (1970)
45. Hoeve, C. A. J.: J. Polym. Sci. Part C. *34*, 1 (1971)
46. Flory, P. J.: Principles of Polymer Chemistry, p. 541, Ithaca, N. Y., Cornell University Press 1953
47. Silberberg, A.: J. Chem. Phys. *48*, 2835 (1968)
48. Scheutjens, J. M. H. M., Fleer, G. J.: J. Phys. Chem. *83*, 1619 (1979)
49. Roe, R-J.: J. Chem. Phys. *60*, 4192 (1974)
50. Scheutjens, J. M. H. M., Fleer, G. J.: J. Phys. Chem. *84*, 178 (1980)
51. Hesselink, F. Th.: J. Colloid Interface Sci. *50*, 606 (1975)
52. De Gennes, P. G.: Rep. Prog. Phys. *32*, 187 (1969)
53. Jones, I. S., Richmond, P.: J. Chem. Soc., Faraday Trans. II *73*, 1062 (1977)
54. De Gennes, P. G.: Scaling Concepts in Polymer Physics, p. 35, Ithaca, N. Y., Cornell University Press 1979
55. Frisch, H. L., Stillinger, F. H.: J. Phys. Chem. *66*, 823 (1962)
56. Hermans, J. J., Overbeek, J. Th. G.: Rec. Trav. Chim. *67*, 761 (1948)
57. Hesselink, F. Th.: J. Colloid Interface Sci. *60*, 448 (1977)
58. Cohen Stuart, M. A., Scheutjens, J. M. H. M., Fleer, G. J.: J. Polymer Sci., Polym. Phys. Ed. *18*, 559 (1980)
59. Varoqui, R., Dejardin, P.: J. Chem. Phys. *66*, 4395 (1977)
60. Passaglia, E., Strombeg, R. R., Kurger, J.: Ellipsometry in the Measurement of Surfaces and Thin Films, NBS Miscellaneous Publication 256, Washington, D. C., Superintendent of Documents, U. S. Government Printing Office 1964
61. Bashara, N. M., Buckman, A. B., Hall, A. C.: Recent Developments in Ellipsometry, Amsterdam, North-Holland Publ. Co. 1969
62. Bashara, N. M., Azzam, R. M. A.: Ellipsometry, Amsterdam, North-Holland Publ. Co. 1976
63. Peyser, P., Stromberg, R. R.: J. Phys. Chem. *71*, 2066 (1967)
64. Fox, K. K., Robb, I. D., Smith, R.: J. Chem. Soc., Faraday Trans. I. *70*, 1186 (1974)
65. Killmann, E., Eckart, R., Makromol. Chem. *144*, 45 (1971)
66. McCrackin, F. L., Colson, J. P.: Computational techniques for the use of the exact drude equations in reflection problems, in: Ellipsometry in the Measurement of Surfaces and Thin Films (eds.) Passaglia, E., Stromberg, R. R., Kurger, J., p. 61, NBS Miscellaneous Publication 256, Washington, D. C., Superintendent of Documents, U. S. Government Printing Office 1964
67. Takahashi, A. et al.: J. Macromol. Sci. Phys. *B17*, 747 (1980)
68. Stromberg, R. R., Tutas, D. J., Passaglia, E.: J. Phys. Chem. *69*, 3955 (1965)
69. Gebhard, H., Killmann, E.: Angew. Makromol. Chem. *53*, 171 (1976)
70. Takahashi, A. et al.: Macromolecules *13*, 884 (1980)
71. Perkel, R. J., Ullman, R.: J. Polym. Sci. *54*, 127 (1961)
72. Berry, G. C.: J. Chem. Phys. *44*, 4550 (1966)
73. Stromberg, R. R., Smith, L. E.: J. Phys. Chem. *71*, 2470 (1967)
74. Kawaguchi, M., Takahashi, A.: J. Polym. Sci. Polym. Phys. Ed. *18*, 2069 (1980)
75. McIntyre, D. et al.: J. Phys. Chem. *66*, 1932 (1962)

76. Fox, Jr., T. G., Flory, P. J.: J. Am. Chem. Soc. *73*, 1915 (1951)
77. Killmann, E., Wiegand, H-G.: Makromol. Chem. *132*, 239 (1970)
78. Öhrn, O. E.: J. Polym. Sci. *19*, 199 (1956)
79. Öhrn, O. E.: Ark. Kemi, *12*, 397 (1958)
80. Tuijnman, C. A. F., Hermans, J. J.: J. Polym. Sci. *25*, 385 (1957)
81. Rowland, F. W., Eirich, F. R.: J. Polym. Sci., Part A-1, *4*, 2033 and 2401 (1966)
82. Priel, Z., Silberberg, A.: J. Polym. Sci. Polym. Phys. Ed. *16*, 1917 (1978)
83. Gramain, Ph.: Makromol Chem. *176*, 1875 (1975)
84. Gramain, Ph., Myard, Ph.: Macromolecules *14*, 180 (1981)
85. Garvey, M. J., Tadros, Th. F., Vincent, B.: J. Colloid Interface Sci. *49*, 57 (1974)
86. Garvey, M. J., Tadros, Th. F., Vincent, B.: ibid. Sci. *55*, 440 (1976)
87. Cebula, D. J. et al.: Faraday Disc. Chem. Soc. *65*, 76 (1978)
88. Barnett, K. G. et al.: Polymer *22*, 283 (1981)
89. Lyklema, J., Vliet, T. V.: Faraday Disc. Chem. Soc. *65*, 25 (1978)
90. Thies, C., Peyser, P., Ullman, R.: The Determination of polymer structures at a liquid-solid interface by infrared analysis, in: Proceedings of the 4th International Congress on Surface Active Substrates Vol. II, p. 1041, New York, N. Y., Gordon and Breach, Science Publishers 1967
91. Peyser, P., Tutas, D. J., Stromberg, R. R.: J. Polym. Sci. A-1, *5*, 651 (1967)
92. Thies, C.: J. Phys. Chem. *70*, 3783 (1966)
93. Thies, C.: Macromolecules *1*, 335 (1968)
94. Joppien, G. R.: Makromol. Chem. *175*, 1931 (1974)
95. Joppien, G. R.: Makromol. Chem. *176*, 1129 (1975)
96. Dietz, E.: Makromol. Chem. *177*, 2113 (1976)
97. Day, J. C., Robb, I. D.: Polymer *21*, 408 (1980)
98. Korn, M., Killmann, E.: J. Colloid Interface Sci. *76*, 19 (1980)
99. Brebner, K. I., Chahal, R. S., St-Pierre, L. E.: Polymer *21*, 533 (1980); Berner, K. I. et al.: Polymer *22*, 56 (1981)
100. Robb, I. D., Smith, R.: Europ. Polym. J. *10*, 1005 (1974)
101. Robb, I. D., Smith, R.: Polymer *18*, 500 (1977)
102. Linden, C. V., Leemput, R. V.: J. Colloid Interface Sci. *67*, 48(1978)
103. Kawaguchi, M., Hayakawa, K., Takahashi, A.: Polymer J. *12*, 265 (1980)
104. Hair, M. L.: J. Colloid Interface Sci. *59*, 532 (1977)
105. Korn, M., Killmann, E., Eisenlauer, J.: J. Colloid Interface Sci. *76*, 7 (1980)
106. Kawaguchi, M., Takahashi, A.: J. Polym. Sci., Polymer Phys. Ed. *18*, 943 (1980)
107. Zimm, B. H., Stockmayer, W. H.: J. Chem. Phys. *17*, 1301 (1949)
108. Takahashi, A. Yamashita, Y.: Morphology, crystallization and surface properties of styrene-tetrahydrofuran block polymers, in: Copolymer, Polyblends and Composites, Platzer, N. A. J. (ed.), p. 267, in: Adv. Chem. Ser., No. 142, Washington, D. C., Am. Chem. Soc. 1975
109. Dawkins, J. V., Taylor, G.: J. Chem. Soc., Faraday Trans. I. *76*, 1263 (1980)
110. Michaels, A. S., Morelos, O.: Ind. Eng. Chem. *47*, 1801 (1955)
111. Schmidt, W., Eirich, F. R.: J. Phys. Chem. *66*, 1907 (1962)
112. Peyser, P., Ullman, R.: Polym. Sci. *A-3*, 3165 (1965)
113. Bartels, T., Arends, J.: J. Polym. Sci., Polym. Chem. Ed. *19*, 127 (1981)
114. Takahashi, A., Kawaguchi, M., Kato, T.: Adsorption of polyelectrolyte studied by ellipsometry, in: Adhesion and Adsorption of Polymers, Part B (ed.), Lee, L-H., p. 729, New York, N. Y., Plenum Press 1980
115. Frommer, M. A., Miller, I. R.: J. Phys. Chem. *72*, 1834 (1968)
116. Nagasawa, M., Izumi, M., Kagawa, I.: J. Polym. Sci. *37*, 375 (1959)
117. Nagasawa, M. et al.: J. Polym. Sci. *38*, 213 (1959)
118. Alexandrowicz, Z., Katchalsky, A.: J. Polym. Sci., *A-1*, 3231 (1963)
119. Devore, D. I., Manning, G. S.: Limiting laws for equilibrium and transport properties of polyelectrolyte solutions, in: Polyelectrolytes (ed.) Sélégny, E., p. 9., Dordrecht-Holland, D. Reidel Publ. Co. 1974
120. Takahashi, A., Nagasawa, M.: J. Am. Chem. Soc. *86*, 543 (1964)
121. Noda, I., Tsuge, T., Nagasawa, M.: J. Phys. Chem. *74*, 710 (1970)

122. Nagasawa, M., Takahashi, A.: Light scattering from polyelectrolyte solutions, in: Light Scattering from Polymer Solutions (ed.) Huglin, M. B., p. 671, New York, N. Y., Academic Press 1972
123. Fixman, M., Skolnick, J.: Macromolecules *11*, 863 (1978)
124. Pefferkorn, E., Dejardin, P., Varoqui, R.: J. Colloid Interface Sci. *63*, 353 (1978)
125. Kawaguchi, M. et al.: Polym. J. *12*, 849 (1980)
126. Maternaghan, T. J., Ottewill, R. H.: J. Photo Sci. *22*, 279 (1974)
127. Kudish, A.: Dissertation, Polytechnic Institute of Brooklyn, 1972
128. Chao, H. Y.: Dissertation, Polytechnic Institute of New York, 1974

Received December 14, 1981
H. Fujita (editor)

Polymer Elasticity
Discrete and Continuum Models

Robert G. C. Arridge and Peter J. Barham

H. H. Wills Physics Laboratory, University of Bristol, Royal Fort, Tyndall Avenue, Bristol BS 8 1 TL, England

Dedicated to Professor Manfred Gordon, University of Essex, with best wishes for his 65th birthday.

There is at present available in the literature on polymers and on materials science a wealth of information regarding measurements of mechanical properties. These properties are dependent upon many relevant physical parameters and most measurements take this into account. There is also available a great deal of information regarding the relations between molecular structure and macroscopic physical properties and many calculations have been made. The bridge between these two extremes (the macro and the micro) is constructed primarily by the use of models of structure.

Composites theory, which has developed from classical elasticity, combined with modelling techniques may point the way forward to a complete theory of the behaviour of polymers. However, it is clear from the literature that many experimentalists do not appreciate the niceties of the mathematical theories of elasticity and of continuum mechanics, nor, in some cases, the inaccuracies inherent in their experimental methods, while nearly all theorists have no conception of the problems encountered by the experimentalist when dealing with real materials and samples of finite size. We have therefore attempted in this review to bring theory and experiment closer together by highlighting some of the problems both of the theoretician and of the experimentalist.

After an introductory chapter we review in Chap. 2 the classical definition of stress, strain and modulus and summarize the commonly used solutions of the equations of elasticity. In Chap. 3 we show how these classical solutions are applied to various test methods and comment on the problems imposed by specimen size, shape and alignment and also by the methods by which loads are applied. In Chap. 4 we discuss non-homogeneous materials and the theories relating to them, pressing the analogies with composites and the value of the concept of the representative volume element (RVE). Chapter 5 is devoted to a discussion of the RVE for crystalline and non-crystalline polymers and scale effects in testing. In Chap. 6 we discuss the methods so far available for calculating the elastic properties of polymers and the relevance of scale effects in this context.

1 Introduction 69

2 Classical Theory 71
2.1 Linear Elasticity – A Brief Outline 72
2.1.1 The Relation Between Stress and Strain 73
2.2 Some Solutions of Equations of Elasticity 74
2.2.1 Uniform Bar Under Uniaxial Tension 74
2.2.2 Torsion of a Uniform Cylinder 75
2.2.3 Bending of a Uniform Bar 77
2.2.4 Indentation by a Sphere 78

Advances in Polymer Science 46

2.3 Viscoelasticity 78
2.4 Summary 81

3 Test Methods 81
3.1 Tensile Tests 82
3.1.1 Description of the Tests 82
3.1.2 The Apparatus 82
3.1.3 Interpretation of Data 83
3.2 Torsion Tests 83
3.2.1 Description of the Tests 83
3.2.2 The Apparatus 84
3.2.3 Interpretation of Data 84
3.3 Beam Tests 84
3.3.1 Description of the Tests 84
3.3.2 The Apparatus Used 84
3.3.3 Interpretation of the Results 85
3.4 Indentation by a Ball 85
3.5 Limitations of the Tests, Especially when Applied to Viscoelastic Non-Linear Inhomogeneous Anisotropic Materials 85
3.6 Material Problems 86
3.6.1 Non-Linearity 86
3.6.2 Inhomogeneity 86
3.6.3 Anisotropy 87
3.7 Geometrical Problems 87
3.7.1 End-Effects Due to Gripping 87
3.7.2 Effects of Specimen Alignment 90
3.7.3 Uniformity of Specimen 92
3.8 Thermodynamic Effects 93

4 Inhomogeneous Materials 94
4.1 Scale 95
4.2 Ordered Heterogeneous Materials with Phases Considered as Continua. The Unit Cell Approach 98
4.3 Bounds on Elastic Constants 100
4.4 Self-Consistent Method 102
4.5 The Self-Consistent Theory Involves Three Stages 104

5 The Representative Volume Element for Solid Polymers 107
5.1 Amorphous Polymers 108
5.2 Amorphous Copolymers 109
5.3 Semi-Crystalline Polymers 110

6 Calculation of the Elastic Constants of Polymers 111
6.1 The Microstructure of Polymers 114

7 Conclusion 115

8 References 115

1 Introduction

1.1 In his "Treatise on the Mathematical Theory of Elasticity" Art. 63, Love[1)] states that "the object of experimental investigations of the behaviour of elastic bodies may be said to be the discovery of numerical relations between the quantities that can be measured... to serve as a basis for the inductive determination of the form of the intrinsic energy function". He goes on to state that this object has not been achieved, except in the case of gases that are far removed from critical states. Love's statement applies only to elastic bodies, that is ones in which a stored elastic energy (strain energy) function can be defined. In the thermodynamics of elasticity this function, W, is identifiable with the intrinsic, or internal energy under adiabatic conditions (dS = 07) or with the Helmholtz free energy A under isothermal conditions (T constant).

In non-polymeric materials the entropy change on deformation is minimal so that the intrinsic and stored elastic energies are the same at least for rapidly occurring events – but in polymers not only may the entropy contribution predominate but for large strains in rubbers the internal energy term is nearly negligible (but not at small strains where it may amount to 20% of the free energy).

In classical elasticity (small strains) W is a quadratic function of the coefficients of infinitesimal strain e_{ij}, whereas in large strain elasticity the relationship is not quadratic and W is then expressed as a polynomial in the strain coefficients or, as is usual in continuum mechanics, as a polynomial in the nine components of the deformation gradient tensor $\frac{\partial x_i}{\partial X_j}$.

In this expression the x_i, X_j are coordinates in the deformed and undeformed axes respectively.

W is, in general, also a function of temperature which itself may be a scalar function of position.

In all the above, however, we have assumed the material body to be homogeneous in composition so that the coefficients of the polynomial expansion of W in the e_{ij} $\left(\text{or } \frac{\partial x_i}{\partial X_j}\right)$ will be independent of position. This is not necessarily the case for two possible reasons. First, as Green and Adkins[2)] point out, the material may possess curvilinear aelotropy, that is the elastic properties at any point may be the same provided that they are referred to local axes which have been suitably rotated. Second, and more importantly, the material may be physically heterogeneous as, for example, a composite made from two or more separate constituents or phases.

In such cases W will be a function of position as well as of temperature and the coordinates of the deformation gradient tensor. Finally, most materials, in particular polymers, are anelastic. Energy is dissipated in them during a deformation and the stored energy function W cannot be defined. It is still of value, however, to consider ideal materials in which W does exist and to seek its form since such ideal materials may approximate quite closely to the real ones.

1.2 There are, however, two other views which may be taken as to the purpose of experiments on the mechanical properties of solids: those of the user and those of the materials scientist. First, the practical user of solids as load-bearing materials requires to know the relationships between load, displacement and time in order to design a structure which may be, for example, a bridge, an engine, a pipe or a woven garment. In all these applications the usual concern of the engineer is to seek numerical quantities such as modulus, elongation to yield and to break, hardness, yield stress and so on. These are available in tabulated form, the result of physical testing in many laboratories, using standard methods, over a number of years. With the exception of rubber nearly all engineering materials are treated as classically elastic, obeying Hooke's law to yield or to break, depending on the material. In most applications isotropy is assumed and the elastic problems encountered solved either by analytical methods or by the use of numerical techniques such as the finite element method. The relations that exist between the mechanical properties of a material and its microscopic and molecular structure are not, in general, the concern of the user of materials but rather of the materials scientist – metallurgist, physicist or chemist. (Nor it may be said, are the niceties of continuum mechanics yet the concern of the materials user. The information which the user requires is derived, as stated earlier, from standard tests analysed by means of classical elasticity).

A fundamental point to be borne in mind when test methods are being devised, and when the results from various types of test are being considered, is that what is being tested is a bulk property and that in deducing this property from the results of the test a theory has to be used which applies to all the material in the test. This is not always taken into account, (although in many cases the finer points of the theory become irrelevant). For example, consider a simple tensile test in which a flat strip is gripped at each end and strained at a constant rate.

It is assumed that the grips do not contribute appreciably to the deformation of the specimen if its length to diameter ratio is large enough. For accurate work surface strain gauges are used to measure the strain at the centre of the specimen, yet in highly anisotropic materials the strain may not be uniform over the cross section so that surface strain may still give an erroneous measure. In torsion tests and in bending tests an additional problem arises in that the strain is by definition non-uniform. Where Hooke's law holds the analysis of the deformation is straightforward but if non-linearity is present, as in most polymers,then the simple St. Venant theory of torsion and the Bernoulli-Euler beam theory for bending no longer apply. Furthermore, in polymers, time and temperature are important variables and in dynamic as in static (creep) testing they need to be considered. Lastly, polymers are not homogeneous, so that the assumption made that bulk properties are being tested needs to be qualified as to the scale on which the word "bulk" is being used. These points are discussed in detail below.

1.3 For the materials scientist the measurement of mechanical properties provides information about internal structure and such measurements afford an opportunity to test theories of structure. In consequence a range of tests has been developed by materials scientists which supplement the, usually simpler, engineering tests and which in many cases provide information of no obvious value to the engineer. There are, however, many *ad hoc* tests used by the practical man which provide information, applicable by the materials scientist, but often in a confusing or complicated way.

An example is the heat distortion test (ASTM D 648-45 T), which is reproducible and of practical use but yields little information of value to the materials scientist. The reason for this lies in the fact that the quantity measured in the test is a function of many variables whereas the materials scientist usually seeks measures which are functions of one variable alone (frequency, temperature etc.). The same is true of a number of practical tests which, though reproducible, are more complicated than the specially developed laboratory tests. It is in fact difficult to design test methods for materials which do yield fundamental information. This is particularly so when highly anisotropic materials are considered: this is discussed in Chap. 3.

Tests which yield structural information to the materials scientist are:

a) Direct mechanical methods:
   Creep tests
   Stress relaxation tests

Measurement of internal friction

These yield a "spectrum of relaxation times", according to theories of linear viscoelasticity.

b) Indirect mechanical methods:
   Ultrasonic wave propagation
   Laser Raman scattering
   Neutron inelastic scattering
   Strain birefringence and other optical methods
   X-ray measurement of strain.

The methods quoted under a) above give bulk information only although they may be used in conjunction with composite models to test theories of the microstructure. The methods under b) are more closely related to structural elements. It is an interesting fact that even at the atomic level, displacement and therefore strain can be measured by several means (although average values are of course obtained) yet loads are only measurable in terms of the secondary effects they produce, for example elastic or inelastic displacements, strain-related optical effects or electronic transitions detectable by optical or infrared spectroscopy. The problem of load transference in a polymer is of great interest, yet very few methods exist by which it may be studied.

Considering these three approaches to the study of elastic materials we have on the one hand measurement of bulk properties using continuum theories and ignoring details of structure, whereas on the other hand properties are used as a tool for the examination of that structure. The bridge between these two extremes lies in theories of structure aimed at explaining the bulk properties in molecular terms. In some cases the bridge is nearly complete, in others partly assembled or totally absent.

2 Classical Theory

The purpose of this chapter is to remind the reader of the basis of the theory of elasticity, to outline some of its principal results and to discuss to what extent the classical theory can be applied to polymeric systems. We shall begin by reviewing the definitions of stress and strain and the compliance and stiffness matrices for linear elastic bodies at small strains. We shall then state several important exact solutions of these equations under idealised loading conditions and briefly discuss the changes introduced if realistic loading conditions are considered. We shall then move on to a discussion of viscoelasticity and its application to real materials.

2.1 Linear Elasticity – A Brief Outline

Strain, in the elementary definition, is a non-dimensional measure of deformation (change of length divided by original length). It is formalized using the classical definitions and the language of Cartesian tensors as follows. In the unstrained material the length ds between two adjacent points **r** and **r** + d**r** is given by $ds^2 = dx_i dx_i$, where x_i is the vector **r** in suffix notation, and the summation convention is assumed. Then in the strained material we have similarly $ds'^2 = dx_i' dx_i'$, where $d\mathbf{r}' = d\mathbf{r} + d\mathbf{u}$ and $\mathbf{u} = \mathbf{r}' - \mathbf{r}$ and we refer all quantities to axes in the unstrained material.
Then

$$\begin{aligned} ds'^2 = dx_i' dx_i' &= (dx_i + du_i) \cdot (dx_i + du_i) \\ &= dx_i dx_i + 2\, du_i dx_i + du_i du_i \\ &= ds^2 + 2\, \varepsilon_{ij} dx_i dx_j \end{aligned}$$

where

$$\begin{aligned} \varepsilon_{ij} &= \frac{\partial u_i}{\partial x_j} + \frac{1}{2} \frac{\partial u_k}{\partial x_i} \frac{\partial u_k}{\partial x_j} \\ &= \frac{1}{2} \left(\frac{\partial u_i}{\partial x_j} + \frac{\partial u_j}{\partial x_i} + \frac{\partial u_k}{\partial x_i} \frac{\partial u_k}{\partial x_j} \right) \end{aligned}$$

ε_{ij} is a measure of large strain expressed in terms of the displacement **u** and the unstrained coordinate system. For small strains the quadratic terms are omitted and we have $e_{ij} = \frac{1}{2}\left(\frac{\partial u_i}{\partial x_j} + \frac{\partial u_j}{\partial x_i} \right)$, the strain tensor used in classical infinitesimal elasticity (Love[12)], Sokolnikoff[3)]). The strain tensor is symmetric i.e. $e_{ij} = e_{ji}$ so that there are only 6 independent strains. It is worth emphasizing at this point that the tensors ε_{ij} and e_{ij} are mathematical quantities requiring differentiability of the displacement vector **u**, a property not realisable in practice in a real solid. The precise meaning of strain at a point in a body therefore requires careful consideration particularly when, as in polymers, relationships between molecular structure and macroscopic properties are being sought.

In the definition of stress similar considerations of continuity and differentiability arise, for the classical definition of stress at a point derives from taking the limit of vanishing area when defining stress as force per unit area.

The tensor of stress, σ_{ij}, has the meaning of the force in direction j on an infinitesimal area with normal in the direction i and is again a symmetric tensor with 6 independent components. In classical elasticity only the force resultant at any point is considered, the couple that must also exist is assumed to be negligible by comparison. However, in polar field theories of elasticity, couple stresses are considered and additional equations of equilibirum required. Classically however only the equation of stress equilibrium

$$\frac{\partial \sigma_{ji}}{\partial x_j} + F_i = 0$$

where the F_i are body forces, is required.

2.1.1 The Relation Between Stress and Strain

In classical infinitesimal elasticity this is expressed by Hooke's law

$$\sigma_{ij} = C_{ijk\ell} e_{k\ell} , \quad \text{or} \quad e_{ij} = s_{ijk\ell} \sigma_{k\ell}$$

where $c_{ijk\ell}$ and its inverse $s_{ijk\ell}$ are fourth rank tensors.

Hooke's law relates stress (or strain) at a point to strain (or stress) at the same point and the structure of classical elasticity (see e.g. Love, Sokolnikoff) is built upon this linear relation. There are other relationships possible. One, as outlined above (see e.g. Green and Adkins) involves the large strain tensor ε_{ij} which does not bear a simple relationship to the stress tensor, another involves the newer concepts of micropolar and micromorphic elasticity in which not only the stress but also the couple at a point must be related to the local variations of displacement and rotation. A third, which may prove to be very relevant to polymers, derives from non-local field theories in which not only the strain (or displacement) at a point but also that in the *neighbourhood* of the point needs to be taken into account. In polymers, where the chain is so much stiffer along its axis than any interchain stiffness (consequent upon the vastly different forces along and between chains) the displacement at any point is quite likely to be influenced by forces on chains some distance away.

Comprehensive accounts of polar field theories (micropolar and micromorphic) and of non-local field theories are given in the book edited by Eringen[4]. We return to Hooke's law, $\sigma_{ij} = C_{ijk\ell} e_{k\ell}$.

Since there are only 6 independent components of stress and strain there are 36 components to **S**, the compliance tensor and **C**, the stiffness tensor. These 36 components may be further reduced using thermodynamic arguments so that there are 21 independent constants for triclinic symmetry, 13 for monoclinic, 7 for tetragonal, 5 for hexagonal, 3 for cubic and 2 for isotropic materials. It is consequently more convenient to use the simplified notation of Voigt where:

e_{11}, σ_{11}	are written as e_1, σ_1
e_{22}, σ_{22}	e_2, σ_2
e_{33}, σ_{33}	e_3, σ_3
$2e_{23}, \sigma_{23}$	e_4, σ_4
$2e_{31}, \sigma_{31}$	e_5, σ_5
$2e_{12}, \sigma_{12}$	e_6, σ_6

The contracted forms for Hooke's law are now in the form of matrix equations:

$$\sigma_i = C_{ij} e_j$$
$$e_i = S_{ij} \sigma_j \quad \text{(summing over j = 1, ... 6)}$$
$$\text{with } \{S\} = \{C\}^{-1}$$

We may then write the arrays for the elastic constants for various symmetries, the two most useful being hexagonal (also transverse isotropy as in fibre symmetry) and isotropic. Hexagonal gives:

$$\begin{array}{cccccc} C_{11} & C_{12} & C_{13} & & & \\ & C_{11} & C_{13} & & & \\ & & C_{33} & & & \\ & & & C_{44} & & \\ & & & & C_{44} & \\ & & & & & 1/2(C_{11} - C_{12}) \end{array}$$

Isotropic gives:

$$\begin{array}{cccccc} C_{11} & C_{12} & C_{12} & & & \\ & C_{11} & C_{12} & & & \\ & & C_{11} & & & \\ & & & 1/2(C_{11} - C_{12}) & & \\ & & & & 1/2(C_{11} - C_{12}) & \\ & & & & & 1/2(C_{11} - C_{12}) \end{array} .$$

The matrix C can be inverted to give the corresponding matrix S, with the relationships to the conventional elastic constants E, G, ν as shown.

$$\begin{array}{cccccc} S_{11} & S_{12} & S_{12} & & & \\ & S_{11} & S_{12} & & & \\ & & S_{11} & & & \\ & & & 2(S_{11} - S_{12}) & & \\ & & & & 2(S_{11} - S_{12}) & \\ & & & & & 2(S_{11} - S_{12}) \end{array} = \begin{array}{cccccc} 1/E & -\nu/E & -\nu/E & & & \\ & 1/E & -\nu/E & & & \\ & & 1/E & & & \\ & & & 1/G & & \\ & & & & 1/G & \\ & & & & & 1/G \end{array}$$

where E is the Young modulus, ν Poisson's ratio and G the shear modulus:

so that $G = \dfrac{E}{2(1 + \nu)}$ for isotropic bodies.

For lower symmetries there is more than one Young Modulus and shear modulus and Poisson's ratio, and their interrelation is not so simple.

2.2 Some Solutions of Equations of Elasticity

2.2.1 Uniform Bar Under Uniaxial Tension

This is the simplest case, and for the idealised loading of uniform tractions on the ends of the bar we obtain the simple relation $\sigma = eE$, where σ is the stress on the cross-section, e the strain and E the Young modulus in the direction of strain.

However, in practice, as we shall show in the next chapter it is not possible to apply such a uniform traction on the ends of a bar and the stress must diffuse into the bar. This causes an end effect. This was first examined for isotropic beams by St. Venant[5)] who considered the cases of uniaxial tension, of torsion and of bending. He proposed that the assumed uniform distribution of forces over any section within the specimen was a limiting state to which the forces in the real specimen approached, the further from the extremities of the specimen. This proposal has become known as St. Venant's principle and is often interpreted as implying that local eccentricities are not felt at distances greater than the largest linear dimension of the area over which the forces are distributed. St. Venant in fact made no calculations of the "end-effect" but in specific cases (e.g. circular cylinder) exact solutions of the elastic problem can be found and a "decay length" proposed. The subject has a long history commencing with Pochhammer[6] Chree[7] and Filon[8] and persisting to the present day (Sokolnikoff[3] § 28, Lur'e[9], Toupin[10], Horgan[11] Folkes and Arridge[12], Arridge et al.[13], Fama[14], Vendhan and Archer[15]). Toupin[10] in discussing St. Venant's principle has pointed out "if one can construct, or is willing to construct, solutions there is no need for the principle".

Exact solutions are, however, not easy to find nor are they usually manageable by non-mathematicians, and in consequence St. Venant's principle has led to a rule of thumb that if a specimen is about 10 diameters in length the stress under any form of end loading can be considered as uniform. There are several objections to the rule of thumb being taken as general practice.

1) It applies, if at all, only to *isotropic* materials of *regular* shape e.g. circular or rectangular cylinders. It is considerably in error for example for samples of dog bone cross-section (Toupin[10]).

2) It is not valid for anisotropic materials where the "decay length" referred to above is not of the order of one diameter, d, even in regular cylinders but rather of order $d\sqrt{E/G}$ where E, G are the longitudinal Young's and shear moduli respectively. (Horgan[11], Arridge and Folkes[16] Arridge et al.[13]).

3) It is not valid for a composite material in which the lateral dimension is comparable with the dimensions of the separate phases, for one then has an interconnected series of small samples.

In summary, very great care is needed in designing experiments to measure mechanical properties under uniaxial tension, particularly when the specimen is highly anisotropic and/or is a composite with a representative volume element (see Chap. 4) of dimensions approaching those of the sample.

We also consider end effects in the next chapter.

2.2.2 *Torsion of a Uniform Cylinder*

It is possible to write a general solution, in terms of the applied torque M, the relevant longitudinal shear modulus, G, the twist per unit length θ, and a form factor, F, thus

$$M = GF\theta \tag{2.1}$$

F depends on both the shape and on the symmetry of the specimen. For isotropic or transversely isotropic materials (e.g. hexagonal symmetry)

$F = \dfrac{\pi a^4}{2}$ circular rod twisted about centre line

$F = \dfrac{\pi(a^4 - b^4)}{2}$ circular tube twisted about centre line

$F = ab^3 f(a/b)$ rectangular prism $a > b$ twisted about centres of rectangular cross-section

$$f(a/b) = \frac{32}{\pi^4}\left\{\sum_{k=1,3,5\ldots} \frac{1}{k^4}\left(1 - \frac{2\,b/a}{k\pi}\tanh\frac{k\pi}{2\,b/a}\right)\right\}$$

These solutions, derived by St. Venant[5], apply only to small values of θ and to isotropic materials. A basic assumption of the St. Venant theory is that the displacement **u** is given by

$$\mathbf{u} = \theta\,(-x_2x_3,\ x_1x_3,\ f\,(x_1x_2))$$

where the axis of the twisted cylinder lies along the 3-direction and f is a function of x_1, x_2 only, determining the warping of the cross-section.

It is shown in works on large strain elasticity (Green and Adkins[2]) that when θ is large there is a normal stress term proportional to θ^2.

For anisotropic materials torsion is discussed in the books by Love, Lekhnitskii[17] and Hearmon[18]. The torque M now depends not upon one elastic constant only, as in the isotropic case, but upon two. This makes the determination of shear modulus by a torsion test a difficult task and requires careful experimentation. Early work on this for polymers was done by Raumann[19], by Ladizesky and Ward[20] and by Arridge and Folkes[16].

If a circular cylinder of orthotropic symmetry is used with axis along 3, the apparent shear modulus to insert in Eq. (2.1) above is $G = 2/(S_{44} + S_{55})$ so that for this geometry a unique value for S_{44} or S_{55} is unobtainable.

If a rectangular cross section is used, as in the studies cited above, the torque M is given by

$$M = C_{55}ab^3F(c)\theta$$

with

$$F(c) = \frac{1}{3}\left\{1 - \frac{192}{c\pi^5}\sum_{j=0}^{\infty}\frac{1}{(2j+1)^5}\tanh\left[\left(\frac{2j+1}{2}\right)\pi c\right]\right\}$$

and $c = a/(\mu b)$, $\mu = (C_{55}/C_{44})^{1/2}$.

In the above expression one must be careful to choose the dimensions a, b to lie in the 1- and 2-directions respectively, the cylinder axis lying along the 3-direction.

By performing experiments with different values for the aspect ratio a/b of the cross-section it is possible to estimate the elastic constants C_{44}, C_{55} and C_{66} but precise values are not obtainable except by considerable labour.

There is, as in the tensile case, a further complication from end effects which may be very considerable in the case of highly anisotropic specimens.

First, if the specimen is subjected to axial stress σ as well as the twist then as Biot[21] has shown, Eq. (2.1) needs to be modified so that $M = M' - \sigma I$, where M' is the applied torque and I the second moment of area of the cross section with respect to the twist axis. Second, the St. Venant principle needs to be modified, as shown above, if the material is highly anisotropic, and very high length to diameter ratios may be necessary (Folkes and Arridge[12]).

This will be discussed in the following chapter.

2.2.3 *Bending of a Uniform Bar*

This is discussed in standard tests on elasticity, and the anisotropic case in Lekhnitskii[17]. The simplest assumption is that of the Bernoulli-Euler theory in which plane sections remain plane and Poisson contraction is ignored. Young's modulus measured by the deflection of a simply supported beam under three point bending is given by

$$E = \frac{W\ell^3}{48\,I\delta}$$

where W is the applied load, ℓ the length of the beam between supports, I the second moment of area of its cross section and δ its mid-point deflection.

If the beam is bent by couples M then Bernoulli-Euler theory gives the relation

$$\frac{EI}{R} = M$$

for the relation between radius of curvature R and moment M, and approximating $\frac{1}{R}$ by $\frac{d^2y}{dx^2}$ the deflection y at any point x of the beam may be found and E derived by its measurement.

More exact theories for the bending of a beam which do not make the assumptions implicit in the above formulae may be found in the literature. There are, again, problems with end effects, now exaggerated because of the number of points (three or four) required to apply the forces for bending.

St. Venant touched on these in his original work and the validity of St. Venant's principle is normally assumed in tests using beams. Again, however, it should be stressed that in precise work the effects of the points of support should be assessed, since St. Venant's principle is not quantitative nor, as we have pointed out, is it valid in its conventionally stated form, when high anisotropy is present.

The particular case of a material reinforced by inextensible rods or fibres has led to the methods of fibre kinematics of Pipkin and Rogers[22] and of Spencer[23].

The transverse vibrations of a beam are used in the test method known as the vibrating reed and the solution used is that of the differential equation (see e.g. Thomson[24])

$$\frac{k^2E}{\varrho}\frac{\partial^4 y}{\partial x^4} + \frac{\partial^2 y}{\partial t^2} = 0 \tag{2.2}$$

where k is the radius of gyration and ϱ the density.

The solution is

$$y = (A\cosh mx + B\sinh mx + C\cos mx + D\sin mx)\cos\omega t$$

with $m = \left\{\frac{\varrho\omega^2}{Ek^2}\right\}^{1/4}$, $\omega = 2\pi f$ and

A, B, C, D are arbitrary constants.

If one end is clamped as is usual in the method the beam becomes a cantilever and if we write ℓ for its length the condition cosh $m\ell$ cos $m\ell = -1$ applies leading to the set of values of $(m\ell)_i$

$(m\ell)_1 = 1.875104$
$(m\ell)_2 = 4.69409$
$(m\ell)_3 = 7.85437$
$(m\ell)_4 = 10.9956$

For higher values of i : $(m\ell)_i \simeq \left(\frac{2i-1}{2}\right)\pi$

Young's modulus for mode i is then given by

$$E_i = \frac{\varrho\omega^2\ell^4}{k^4(m\ell)_i^4}$$

Higher harmonics than the fundamental can be detected with the vibrating reed device and the variation of E with frequency studied.

As in the simple 3- or 4-point bending of a beam the vibrating reed device assumes the validity of the differential Eq. (2.2) which is due to Euler. Timoshenko[25)] included both rotary inertia and shear deformation deriving a more exact differential equation which reduces to the Euler equation as a special case. Use of the Timoshenko beam theory for anisotropic materials has been made by Ritchie et al.[26)] who derive a pair of equations for torsion-flexure coupling (which will always occur unless the axis of the beam coincides with the symmetry axis of the anisotropic material).

2.2.4 *Indentation by a Sphere*

This case is of particular interest because, at least in the case of isotropic materials, there is no end effect since the idealised loading conditions can actually be realized in a practical situation. The problem was first considered by Hertz[27)] and is treated in greater detail in Love's[1)] article 138. The general result, applicable for a rigid sphere indenting an elastic body at a plane surface is:

$$G = \frac{3RW}{16(2Rd - d^2)^{3/2}}$$

where G is the shear modulus of the elastic body
R is the radius of the rigid sphere
d is the depth of penetration of the sphere
and W is the weight of the sphere

More complex expressions arise if the sphere is taken to be elastic, so that the above equation should only apply to soft materials indented by very stiff materials.

2.3 Viscoelasticity

In the above consideration of the elastic response of materials it has been asssumed that the stress is a linear function of the strain only. In practice this is not true and the stress also varies with time. The simplest and only practical approach is to consider linear viscoelastic behaviour where we may write:

$$A_0\sigma + \ldots + A_n \frac{d^n\sigma}{dt^n} + A_{n-1} \frac{d^{n-1}\sigma}{dt^{n-1}} + \ldots + A_1 \frac{d\sigma}{dt} =$$

$$E_0 e + E_1 \frac{de}{dt} + \ldots + E_n \frac{d^n e}{dt^n} + \ldots \qquad (2.3)$$

where the A_n and E_n's are constants. Since this is a linear differential equation with constant coefficients (In the general case we may write σ and e as tensors so that the A's and E's are constant tensors) we may take advantage of the fact that any linear combination of any solutions will also be a solution. This property of linear differential equations gives rise to the principle of superposition which is of great use in the theory of viscoelasticity.

In order to describe the material fully we should then need to find all the A's and E's, each of which of course would have up to 21 independent components. This is obviously far too complicated so that we instead consider various simplifications that can be made to Eq. (2.3). Examples are:

$$A_0\sigma = E_0 e \quad \text{– Hookes Law}$$

$$A_0\sigma = E_1 \frac{de}{dt} \quad \text{– Newtonian Viscosity}$$

$$A_0\sigma + A_1 \frac{d\sigma}{dt} = E_1 \frac{de}{dt} \quad \text{– The Maxwell Element}$$

$$A_0\sigma = E_0 e + E_1 \frac{de}{dt} \quad \text{– The Voigt Element}$$

$$\text{and} \quad A_0\sigma + A_1 \frac{d\sigma}{dt} = E_0 e + E_1 \frac{de}{dt} \quad \text{– General linear solid.} \qquad (2.4)$$

The last three are used as models for viscoelastic materials, and are often presented in the form of spring-dashpot models which have the same constitutive equations.

The general linear solid leads to the single relaxation time model; the solution of (2.4) for the case of oscillating strain leads to

$$G' = G_R + \frac{G_U - G_R}{1 + \omega^2\tau^2}$$

$$G'' = \frac{(G_U - G_R)\omega\tau}{1 + \omega^2\tau^2}$$

where G' and G'' are the real and imaginary parts of the particular modulus (or component of stiffness tensor) under consideration. G_U and G_R are the values at infinite and zero frequency respectively – known as unrelaxed and relaxed moduli, ω is the measurement frequency and τ a relaxation time. These models behave somewhat like real polymers, although in practice it is necessary to introduce a spectrum of relaxation times to account for the observed broadness of the real loss peaks. Such a distribution is usually introduced by consideration of either a parallel array of Maxwell or a series array of Voigt elements giving, for the Maxwell elements:

$$\sigma = \sum_i \sigma_i = e_0 \sum_i G_i \exp(-t/\tau_i)$$

or for the Voigt elements:

$$e = \sum_i e_i = \sigma_0 \sum_i \tau_i \{1 - \exp(-t/\tau_i)\}$$

If we assume a continuum of relaxation times then we obtain

$$G(t) = \int_0^\infty H(\tau)\exp-(t/\tau)d\tau$$

$$J(t) = \int_0^\infty L(\tau)\{1 - \exp-(t/\tau)\}d\tau$$

where $H(\tau)$ and $L(\tau)$ are distributions of relaxation and retardation times respectively.

If we express these equations in terms of relaxed and unrelaxed moduli we obtain

$$G(t) = G_R + (G_U - G_R)\int_0^\infty H_N(\tau)\exp(-t/\tau)d\tau$$

where

$$H_N(\tau) = \frac{H(\tau)}{G_U - G_R} \text{ is normalised so that } \int_0^\infty H_N(\tau)d\tau = 1$$

and

$$G'(\omega) = G_R + \omega\int_0^\infty \sin\omega t\, G(\tau)d\tau$$

$$G''(\omega) = \omega\int_0^\infty \cos\omega t\,\{G(\tau) - G_R\}d\tau$$

Thus, we may give a good description of a linear viscoelastic material in terms of relaxed, and unrelaxed elastic constants and a distribution of relaxation times (– this is not necessarily the same distribution for each elastic constant!). These all have to be found from experiments. In general it is possible to find some of the relaxed and unrelaxed elastic constants and to estimate the distribution of relaxation times.

In practice it is often more convenient to vary the temperature of an experiment while keeping the frequency fixed in which case for single relaxation time processes we may write

$$\tau = \tau_0 \exp\left(\frac{\Delta H}{RT}\right)$$

and thus relate the relaxation time to the temperature, through an activation energy ΔH. It is generally assumed that we may do this for a distribution of relaxation times with a constant ΔH, although this assumption is presently being brought into some doubt by McCrum et al.[28]. We may now see that, in order to describe a linear viscoelastic material we need to know relaxed and unrelaxed values for the elastic constants, a distribution of relaxation times and an activation energy. Various methods have been used to determine as many as possible of these but in general only the relaxed and unrelaxed moduli, and occasionally the activation energies, are measured. In making any such measurements it is normally assumed that the equations for linear elastic bodies may be used for viscoelastic bodies simply by introducing complex elastic constants into the solutions of elastic problems (this gives dynamic solutions) or by using operators as in Laplace transform theory. Various forms of this "correspondence principle" have been used in the literature[29–31].

2.4 Summary

The exact solutions of the linear elasticity theory only apply for small strains, and under idealised loading conditions, so that they should at best only be treated as approximations to the real behaviour of materials under test conditions. In order to describe a material fully we need to know all the elastic constants and, in the case of linear viscoelastic materials, relaxed and unrelaxed values of each, a distribution of relaxation times and an activation energy. While for non-linear viscoelastic materials we cannot obtain a full description of the mechanical properties.

If comparisons are to be made between various sets of experimental data great care must accordingly be taken to ensure that the data were obtained using comparable experimental conditions. This highlights the importance of stating the exact experimental conditions used when quoting any mechanical properties.

3 Test Methods

In this chapter we shall describe some mechanical tests in common use to measure "moduli" of polymeric materials.

We shall first simply describe the techniques and the commonly used methods of interpreting and presenting the data. We shall then describe in some detail several of the errors which can, and frequently do arise when using these techniques. All too often the possibility of such errors is overlooked and accordingly we suggest appropriate tests to ensure, as far as possible, that the measured moduli are a true reflection of the material properties.

We do not intend this to be in any way a complete compendium of test methods or of the errors which can arise; we simply restrict ourselves to the tests with which we are reasonably familiar and those errors which we believe are most frequently ignored.

3.1 Tensile Tests

3.1.1 Description of the Tests

Tensile tests involve either stretching a sample and monitoring the load or loading it while monitoring the extension. The simplest test uses a tensile testing machine (e.g. an Instron) where the sample is stretched at a constant rate while the load is measured using a (usually hard) load cell. Variations on this test allow the specimen to be extended at a constant strain rate or to be loaded at a constant load, or stress rate. These latter tests are usually carried out on servo hydraulic machines.

The next two tests falling into this category are tensile creep and stress relaxation. In tensile creep a load is applied "instantaneously" to the specimen at zero time and the extension monitored as a function of time. In stress relaxation an extension is imposed and the load monitored as a function of time.

Finally there are dynamic tensile tests where an oscillatory extension is applied to the sample and the resulting oscillatory load is measured.

All these tests are in common use to measure the tensile stiffness of polymers. For example, tests at constant extension rate are often carried out on an Instron tensile testing machine. Tensile creep is used in many cases while stress relaxation is not so common. Dynamic testing is commonly performed using the "Rheovibron" or other commercial equipment[32)] or home made equipment[33, 34)]

3.1.2 The Apparatus

The apparatus used for simple extension tests are usually commercial tensile testing machines. These employ a cross-head moved by lead screws driven by a powerful motor, capable of a range of speeds. Loads are measured using hard load cells connected to appropriate amplifiers. Extensions are preferably measured by some form of extensometer or strain gauge attached to the specimen or less satisfactorily from the cross-head displacement. This is usually done only when it is not always possible to use an extensometer or strain gauge e.g. if the specimen is particularly delicate or if it is softer than the gauge itself. It is necessary to calibrate the system carefully using a standard (e.g. a steel wire) of known stiffness, similar to that of the specimen. It is also necessary to establish that the specimen does not slip in the grips – this may be achieved by returning the cross-head to its initial position and checking that the same trace is obtained after some (long) resting period. This should be an obvious practice but it is often not done.

Creep and stress relaxation tests are more usually carried out on specially designed equipment although commercial machines are available; typical equipment layouts can be seen in the literature[35)]. In creep experiments the common method for measuring extension is by means of linear variable displacement transformers (LVDT) – loads are usually applied by hanging weights on a lever arm. In stress relaxation tests the load is usually monitored using a hard load cell. In both creep and stress relaxation it is very important to guard against slip in grips.

Dynamic loads are most commonly carried out using commercial equipment (e.g. Rheovibron; details of the many problems associated with the use of the Rheovibron can be found in the article by Wedgewood and Seferis[36)]) but for specific applications apparatus is specially built. Displacement is provided either by an electrical transducer or by a mechanical system (e.g. a Scotch Yoke). The load and grip displacements are usually both measured separately as soft load cells are often used.

3.1.3 *Interpretation of Data*

All the uniaxial tensile loads described above assume that the load is shared uniformly across the sample cross-section and that the sample maintains a constant cross-section (over the measured region at least). The tensile modulus, E, is then taken to be defined as:

$$E = \sigma/e \qquad \begin{aligned}\sigma &= \text{stress}\\ e &= \text{strain}\end{aligned}$$

σ and e being determined from the load and extension. However for polymers E is not a constant but depends on the strain, strain rate, time etc. and we have to define it in more precise ways. For the simple uniaxial tension experiment there are two moduli which can be defined, the secant modulus and the tangent modulus. Both will depend on the strain and extension rate at which they are measured. The tangent modulus is the slope of the tangent to the stress-strain curve at the given strain and extension rate. The secant modulus is the slope of the line from the origin to the stress-strain curve at the given strain. For creep (stress relaxation) experiments the modulus (compliance) is a function of both the time and the load (extension) applied. It is usual to give a curve of E (or e) or (or σ) against time often at more than one stress (extension) level. The modulus obtained at very short times should correspond to the unrelaxed value and that at very long times to the relaxed value. An analysis of the shape of the E, t curve can reveal information about relaxation times (if the material has a sufficiently simple behaviour).

For the dynamic experiments the applied extension is converted to a strain $e = e_0 \sin \omega t$ and the load to a stress $\sigma = \sigma_0 \sin(\omega t + \delta)$, and the resulting data are presented as either $E = (\sigma_0/e_0)$ and $\tan \delta$ or $E' = \sigma_0/e_0 \cos \delta$ and $E'' = \sigma_0/e_0 \sin \delta$. Data are usually presented as a function either of frequency or, more commonly, temperature.

3.2 Torsion Tests

3.2.1 *Description of the Tests*

Torsional tests involve applying a torque to the sample and observing the resulting twist. The simplest is the equivalent to the tensile creep experiment and is known as torsional creep; a torque is applied and subsequent twist monitored. The remaining techniques are the free and driven torsion pendulums. The driven pendulum is equivalent to the dynamic uniaxial tension experiment – an alternating twist or torque is applied and the resulting torque, or twist, is measured. The free pendulum is a resonance method – the specimen is given a small twist and the resulting oscillations and their decay are monitored. This method has the advantage that there is no need to measure torques but the disadvantage that measurements are only obtained at the systems resonant frequency. The most easily interpreted measurements on viscoelastic materials are those where either the frequency or measurement temperature is held constant and the other is varied (see section on viscoelasticity in preceding chapter). This is difficult, but not impossible, to achieve with the free torsion pendulum.

3.2.2 The Apparatus

There are a large number of designs for all types of torsional apparatus, mostly home made. In apparatus for torsional creep the measurement of displacement may be made by a variety of methods e.g. optical lever and spot-following recorder, or electrical sensing. The application of the torque may be made by means of weights or electrically.

Free pendulums come in two main categories, those where the specimen itself supports an inertia arm and those where the end load is balanced so that there is no tensile stress in the specimen, these are known as inverted torsion pendulums. In either case the sensing of twist is done usually either by an optical system or by an electrical one.

3.2.3 Interpretation of Data

In the creep experiment one obtains the twist as a function of time, this is usually converted into a shear modulus figure, using the equations described in the previous chapter. The data are usually presented as curves of G v. time at a variety of temperatures – again at short times we have the unrelaxed modulus and at long times the relaxed modulus. The driven pendulum gives as data both the torque and the twist from which we can obtain G and tan δ, or G' and G''. The data are usually presented in plots against either frequency or temperature. The free pendulum gives data on resonant frequency and decay from which we can deduce G, tan δ or G' and G''. The results are usually presented as plots against temperature.

3.3 Beam Tests

3.3.1 Description of the Tests

Tests using beams involve either the forced bending of a beam resting on two fixed points by a force applied centrally between them (3 point bending) or the resonance of a beam clamped at one end. The bending technique may be performed statically, the applied force is a constant, and the deflection is monitored as a function of time, or a deflection which varies with time may be applied and the resulting force at the deflecting point measured. In this latter test the deflection may either increase with time (analogously to the straightforward tensile tests) or vary sinusoidally with time. The resonance tests (also called vibrating reed tests) involve only finding the resonant frequencies of a vibrating beam and calculating the stiffness.

3.3.2 The Apparatus Used

For bending experiments either 3 or 4 point bending may be used. For static or creep tests the load is usually applied as a weight on a hanger and the deflection measured using an

LVDT, an extensometer or strain gauge, or optically. For tests where the deflection is increased with time a tensile testing machine is convenient. For dynamic tests it is necessary to clamp the specimen at the fixed points, the sinusoidal displacement being applied either by a Scotch yoke (as used e.g. by Gibson et al.[37]) or by an electric transducer, the displacement measuring and load weighing systems are closely similar to the dynamic tensile test. The resonant method, or vibrating reed equipment, usually consists of a bar clamped at one end in a transducer which can be vibrated at a range of frequencies. The displacement of the beam is monitored (usually by an optical system) resonances are found by sweeping through the frequency range to find frequencies where the displacement of the beam is maximised. If frequencies of half power points are also noted it is possible to find tan δ as well as E.

3.3.3 Interpretation of the Results

When considering the bending of a beam and attempting to extract a modulus value one must make several assumptions, the most important being that the modulus in tension is the same as in compression, and is independent of strain (at least for the range of strain involved). The simple Bernoulli-Euler theory is usually used to interpret the data. When performing resonance tests it is particularly useful to find a set of resonances and compare the measured frequency ratios with the theoretical ones given in the previous chapter.

3.4 Indentation by a Ball

This test is not in common use for polymers (except for the hardness of rubbers) but we present it here since it has, as we shall show, several advantages, at least for isotropic materials. The test consists essentially of applying a force over a spherical surface – either by placing a known weight ball on the specimen, or by pushing such a ball in with a greater force and monitoring the displacement, whence G can be calculated as shown in Chap. 2.

3.5 Limitations of the Tests, Especially when Applied to Viscoelastic Non-Linear Inhomogeneous Anisotropic Materials

All the tests we have described above only give reliable answers for linear, elastic, isotropic, homogeneous materials, though even then in some cases errors may arise from carelessness in experimental method. In this section we shall try to describe many of the complications which arise when tests are applied to real polymeric materials, in particular to anisotropic materials, and to show how, if at all, they may be avoided.

It is possible to divide the problems essentially into 3 classes: 1) problems associated with the material – here we must concern ourselves with the applicability of the solutions of the equations of elasticity used to calculated moduli to the actual material which may be non-linear, inhomogeneous and anisotropic as well as viscoelastic. 2) Geometrical problems of the test itself; here we must be concerned with the end effects due to gripping the specimen and with such matters as the effect of small misalignments and 3) Thermodynamic problems, normally it is assumed that tests are isothermal (or adiabatic). Since in practice they are unlikely to be either, this will lead to problems for materials whose moduli are particularly temperature sensitive. Also we should be concerned with control of ambient temperature for such materials.

Of course all these problems are interrelated and the above division is of necessity fairly arbitrary, nevertheless we shall use it to discuss the problems.

3.6 Material Problems

We shall discuss separately the effect of non-linearity, inhomogeneity and anisotropy and consider how they affect individual tests. The problem of the translation of elastic solutions to the viscoelastic case has been mentioned in the previous chapter and will not be treated further.

3.6.1 Non-Linearity

The most obvious problem of non-linearity is the *definition* of a modulus. For a linear viscoelastic material we need to define not only a real and an imaginary modulus but also a spectrum of relaxation times if we are fully to describe the material – although it is more usual to quote either an isochronous modulus or a modulus at a fixed frequency. We must, for a full description of a non-linear material give the moduli (and relaxation times) as a function of strain as well; this will not usually be practicable so we satisfy ourselves by quoting the modulus at a given strain. The question then arises as to whether this should be the tangent modulus $\left(\frac{d\sigma}{de}\right)$ or the secant modulus (σ/e); it is more usual to quote a secant modulus although it would be preferable to quote both.

The problem of definition of modulus applies to all tests. However there is a second problem which applies to those tests where the state of stress (or strain) is not uniform across the material cross-section during the test (i.e. to all beam tests and all torsion tests – except those for thin walled cylinders). In the derivation of the equations to determine moduli it is assumed that the relation between stress and strain is the same everywhere, this is no longer true for a non-linear material. In the beam test one half of the beam is in tension and one half in compression with maximum strains on the surfaces, so that there will be different relations between stress and strain depending on the distance from the neutral plane. For the torsion experiments the strain is zero at the centre of the specimen and increases toward the outside, thus there will be different torque-shear modulus relations for each thin cylindrical shell. Unless the precise variation of all the elastic constants with strain is known it will not be possible to obtain reliable values from beam tests or torsion tests (except for thin walled cylinders).

3.6.2 Inhomogeneity

If the material is inhomogeneous (i.e. there are regions within the specimen of differing mechanical properties) as will be the case for composites or for semi-crystalline polymers, then we must consider whether we may apply the solutions we have already given. All the solutions to the elasticity equations we have used have assumed that the elastic constants are the same everywhere. In an inhomogeneous material this is no longer true. We must therefore ensure that the scale of inhomogeneity is sufficiently small compared with the specimen dimensions that the measured properties are typical of the material as a whole. This is the same as saying that the specimen must be very much larger than the Representative Volume Element (RVE) defined in the next chapter. For example for an isotropic crystalline polymer with a spherulite size $\sim 250\,\mu m$ the RVE will be significantly larger ~ 1 mm say so that the specimen must be larger than 1 mm in all dimensions; this may, for example, make the use of torsion tests difficult particularly if the material is also non-linear so that thin walled tubes are required (with a thin wall thickness of several mm the tube diameter would need to be some tens of centimetres at least).

3.6.3 Anisotropy

The problems due to anisotropy may essentially be divided into two categories. First there is the problem that often the particular solution used to calculate a modulus itself assumes isotropy, e.g. for the torsion pendulum few authors use the full expressions when considering anisotropic materials. These problems may, in general, be overcome by finding solutions for anisotropic materials.

Secondly, there are, as we shall see in the following sections, far larger errors introduced by the specimen geometry when using anisotropic specimens. To give just one example, if we use the solutions for an anisotropic body in torsion we are assuming that the torsion axis is one of the specimen's principal axes; if it is not then the solution is no longer applicable.

3.7 Geometrical Problems

There are three principal problems relating to the specimen, and test geometry. These are the end effect due to gripping the sample, the alignment of the specimen in the test equipment, and the shape of the specimen itself. We shall deal with each of these separately.

3.7.1 End-Effects Due to Gripping

When a specimen is gripped, the stress must be transferred from the grip to the specimen. This transfer takes place across the specimen-grip interface by a shear mechanism. Furthermore the stress must diffuse across the entire specimen cross section, before we reach the situations described by the theoretical solutions. We have already discussed the theoretical background to this problem in the previous chapter; here we shall be concerned with the practical problems it imposes on real measurements. For isotropic materials it is usual to adopt the convenient rule of thumb that specimens should have an aspect ratio of at least 10. We feel it is useful to illustrate the effect of aspect ratio on modulus for isotropic materials. In Fig. 1 we present the "modulus", measured as the

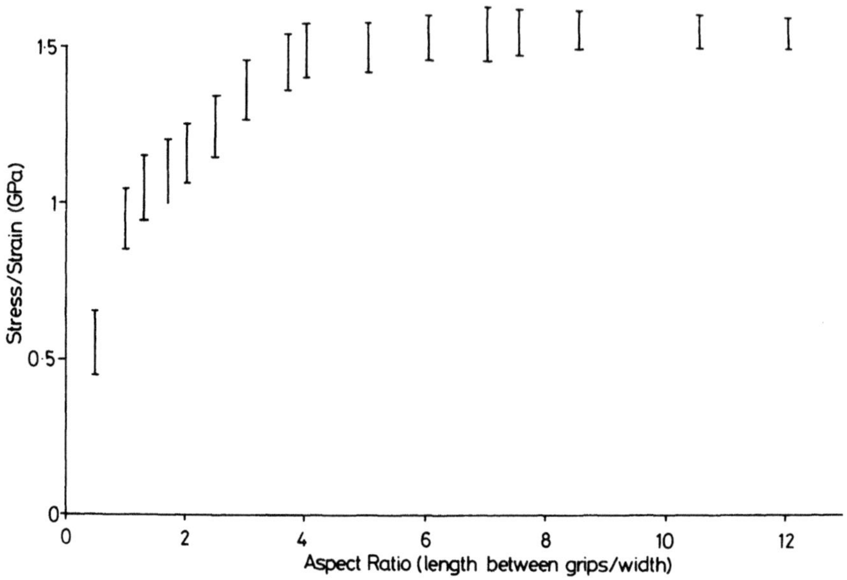

Fig. 1. Apparent secant modulus (stress/strain) in tension at 0.1% strain and at various sample aspect ratios: measured at a strain rate of $10^{-4}s^{-1}$. Isotropic polyethylene sheet at 22 °C (sample cross section 1 mm × 3 mm)

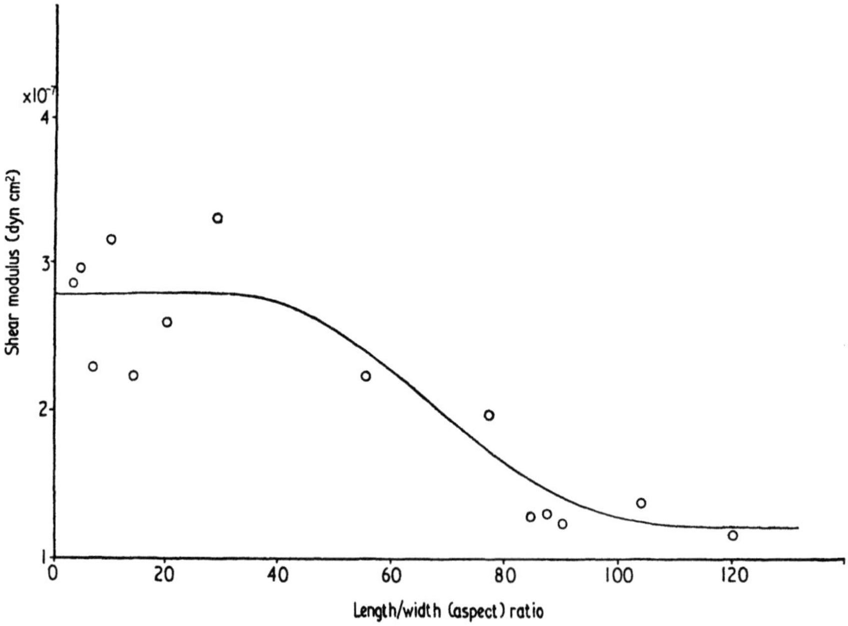

Fig. 2. Apparent shear modulus (from torsion data) of oriented S. B. S copolymer as a function of sample aspect ratio. Curve indicates the general trend of results (From Ref. 12)

secant 0.1% modulus at 10^{-4} s^{-1} strain rate at 22 °C of an isotropic polyethylene sheet, at a variety of sample aspect ratios. It is quite clear that the above rule of thumb is no more than adequate in this case.

For anisotropic materials Horgan has shown[11)] that the stress will not be uniform over a region $\sim\sqrt{E/G}\cdot d$ from the grip (where E is tensile modulus along specimen, G the longitudinal-shear modulus and d the specimen diameter). For anisotropic polymers $\sqrt{E/G}$ can be quite large and the effects strongly pronounced. We shall give a few examples; in torsion the effect of non uniformity is to concentrate the shear stress on the outside of the specimen so that when calculating a shear modulus the material appears to be stiffer than it otherwise would; this has been observed for an oriented SBS (styrene-butadiene-styrene copolymer) sample (Fig. 2)[12)] where we can see that the apparent shear modulus (calculated assuming a uniform state of stress) decreases as the sample aspect ratio increases. In tension, and 3 point bending, the effect causes an underestimation of the modulus at low aspect ratios[13, 37)] this is illustrated for highly oriented polyethylene in Fig. 3. Again we see that at high aspect ratios we obtain a constant value for modulus. As a general rule of thumb samples should have aspect ratios greater than $10\sqrt{E/G}$, which can often mean aspect ratios ~ 100 or more. However it is always advisable to make measurements at several aspect ratios when using anisotropic materials and even isotropic ones – the figure of 10 : 1 quoted above is not sacrosanct. When it is not possible to obtain high aspect ratio samples it is necessary to determine the strain along a central section at least a distance $\sqrt{E/G}$ d from the grip, this may be done using a surface strain gauge of some kind, but care must be taken that the gauge is significantly softer than the specimen; it is also advisable to measure the surface strain at a number of points along the length of the specimen. An example of such a measurement on oriented SBS copolymer[38)] is given in Fig. 4; the strains were measured by monitoring optically the

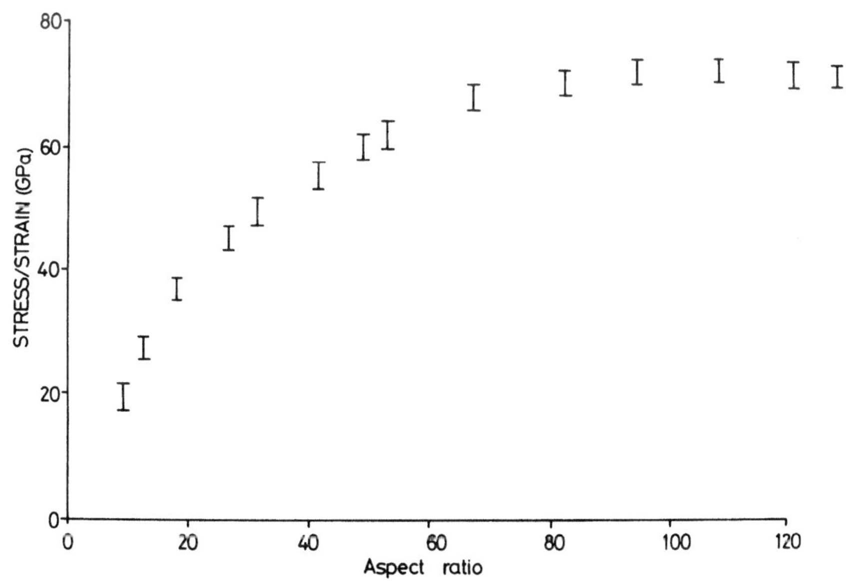

Fig. 3. Apparent secant modulus (stress/strain) in tension at 0.1% strain and at various sample aspect ratios, measured at strain rate of $10^{-4}s^{-1}$. Highly oriented polyethylene (D. R. 28) measured in direction of drawing. (Sample cross section 0.2 mm × 2 mm)

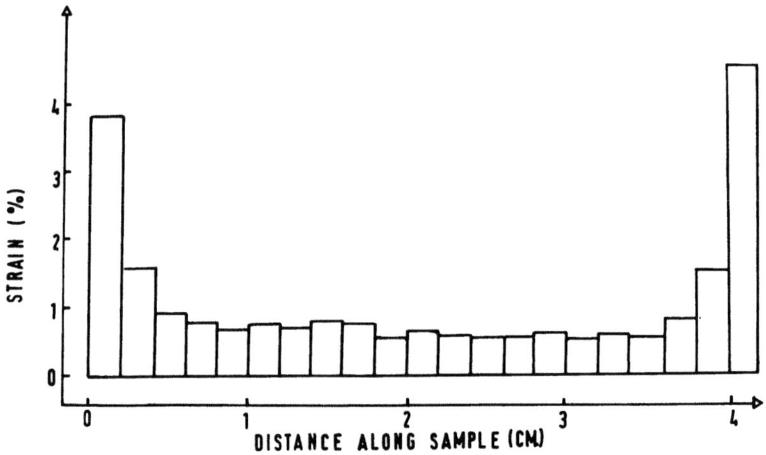

Fig. 4. Distribution of surface strain measured along a specimen of oriented S. B. S. copolymer loaded in tension (From Odell and Keller[38])

distortion of an aluminium grid evaporated on to the specimen surface. Here the effect of the ends can be seen quite clearly, and a sensible value for the modulus can be estimated from the minimum surface strains.

When examining such results it may be useful to visualise the stress using finite element analysis. This has the advantage of giving a dramatic, graphical, representation of non-uniform stress, it also shows that the surface strain may not always be a good guide to the overall internal strain, particularly for very highly anisotropic samples of low aspect ratio. Some examples are shown in Fig. 5 from[16]. As well as plotting the principal stresses in each element of the two dimensional array we have also calculated the modulus from surface strains and from the central strain at various aspect ratios (Fig. 6). It is clear that this surface strain can apparently reach a uniform minimum when the stress is by no means uniform across the entire cross-section.

We can conclude therefore that when measuring the modulus of a specimen one should always ensure that it is long enough that end effects may safely be ignored, and that this is best done by measuring a series of specimens of differing aspect ratio plotting modulus against aspect ratio and extrapolating to the value of the modulus at infinite aspect ratio. Where this is not practicable a good rule of thumb is that the specimen should have an aspect ratio at least greater than $10\sqrt{E/G}$.

3.7.2 Effects of Specimen Alignment

In tensile tests it is assumed that the specimen is aligned along the direction in which the force (or displacement) is applied. This means that the symmetry axis of the specimen (i.e. axis through centre of cross-section) is co-linear with the axis along which the force is applied. If this is not the case then shearing and bending of the specimen will occur. The errors introduced will obviously be most pronounced for anisotropic specimens. An example is given in Fig. 7 where we show the *apparent* modulus of an anisotropic fibre of cross section 0.1×1 mm as a function of misalignment. Large errors can occur for

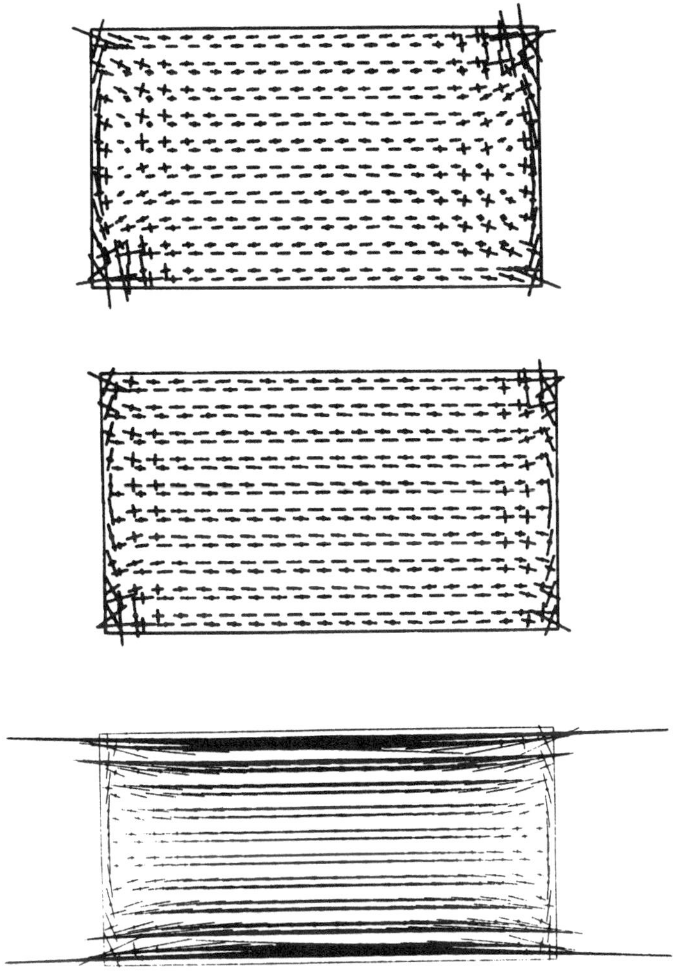

Fig. 5 a–c Principal stress values for rectangular bars loaded along their length in tension by forces applied at the four corners. (Finite element calculations in plane strain). **a** Isotropic material aspect ratio 5; **b** Isotropic material aspect ratio 10; **c** Anisotropic material $(E/G)^{1/2} = 15.1$, aspect ratio 10 (Note: the figures are not drawn to scale). (From Ref. 16)

comparatively small inaccuracies in alignment. In torsion tests the effects are much more pronounced and their presence is relatively easily detected. If the specimen is misaligned in an inverted torsion pendulum then by increasing the end load we tend to reduce the effect of such misalignment, i.e. the force tends to pull the whole system straight, although in such a case the axis of rotation may no longer correspond to a principal axis of the whole system. If we plot the end load against the apparent modulus (or $1/(\text{period})^2$) then we should, for a properly aligned specimen find a linear relationship as shown in the previous chapter, the period decreasing as the load increases. However, when the specimen is misaligned this is no longer the case, instead the period at first increases dramatically with end load. This is illustrated for a variety of deliberately misaligned specimens in Fig. 8. It is therefore wise always to check that $^1/_{T^2}$ rises linearly with end load, thus

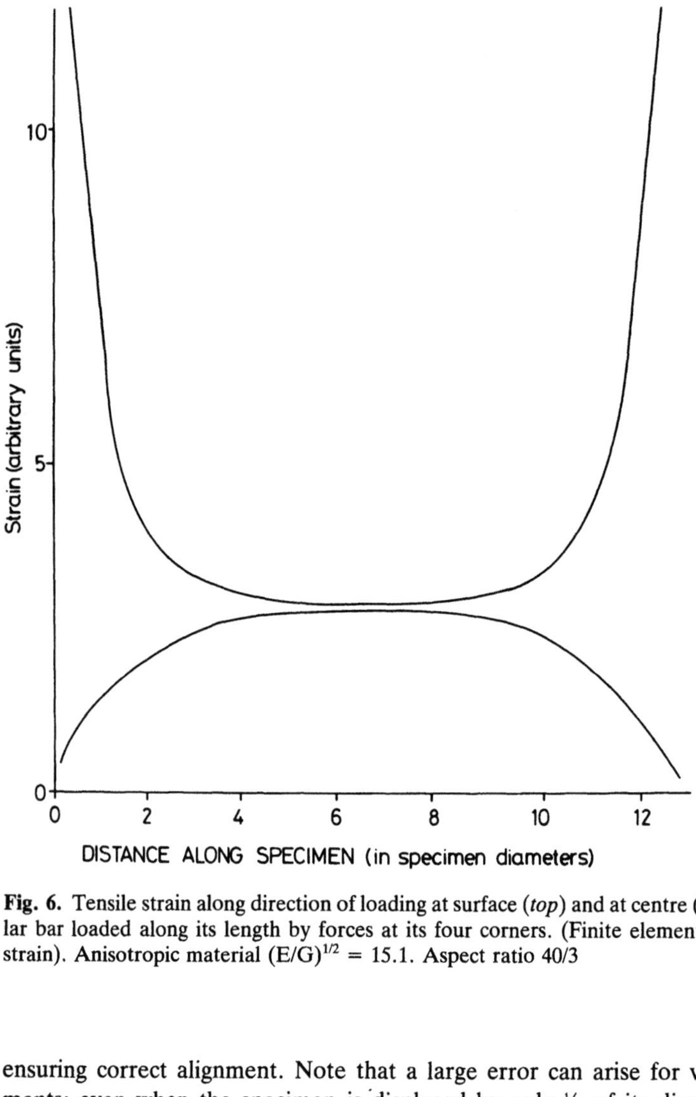

Fig. 6. Tensile strain along direction of loading at surface (*top*) and at centre (*bottom*) in a rectangular bar loaded along its length by forces at its four corners. (Finite element calculations in plane strain). Anisotropic material $(E/G)^{1/2} = 15.1$. Aspect ratio 40/3

ensuring correct alignment. Note that a large error can arise for very small misalignments; even when the specimen is displaced by only 1/10 of its diameter the measured "shear modulus" can be in error by more than 50%!

3.7.3 Uniformity of Specimen

If the specimen does not have a uniform cross-section along its entire length then any moduli measured on it will be, to some extent, in error. Sometimes it may not be practicable to produce completely uniform specimens. In such cases it may be possible to measure a series of different specimen lengths, so that the shortest is least likely to be non-uniform. However, great care must be taken to avoid any introduction of end-effects by using short specimens.

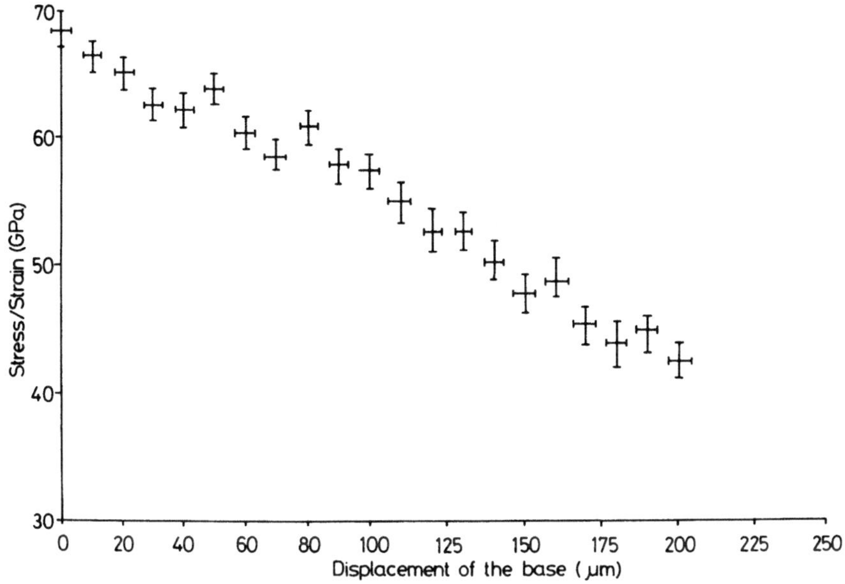

Fig. 7. Apparent secant modulus (stress/strain) in tension at 0.1% strain measured as specimen is deliberately misaligned by displacing one end of the specimen in a direction parallel to its thinnest dimension. Strain rate $10^{-4}s^{-1}$. Temperature 22 °C. Highly oriented polyethylene (D. R. 28). (Sample dimensions 0.1 mm × 2 mm × 300 mm)

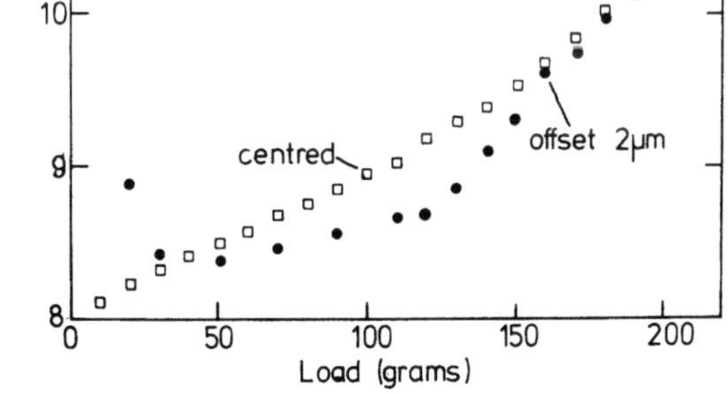

Fig. 8. Apparent torsional modulus (period^{-2} in arbitrary units) measured at various end loads. □ Specimen correctly aligned ● Specimen displaced at one end by 2 microns, parallel to its thin dimension.
× Specimen displaced at one end by 10 microns.
Highly oriented polyethylene (D. R. 29). Temperature 20 °C. (Specimen dimensions 0.1 mm × 2 mm × 100 mm)

3.8 Thermodynamic Effects

Most polymers have very strong temperature dependence of modulus, especially so when they are close to a relaxation, so that small changes in temperature during an experiment can cause large changes in measured moduli. Furthermore polymers hold a memory of the previous mechanical and thermal history. For this reason it is essential that the specimen has been left at the measurement temperature and in the initial state of stress for sufficiently long that all past history has been forgotten. Cooper and McCrum[39] found, for example, that it is necessary to hold specimens for several hours at the measurement temperature before reproducible measurements can be obtained.

When one considers the effect of variation of temperature during a test there are two distinct cases. First in long term tests such as creep it is important that the temperature control is very good ($\pm$ 0.1°); naturally the precision of temperature control becomes more important close to relaxation temperature. Secondly in short term tests we must decide whether the experiment is conducted isothermally, or adiabatically. The moduli will of course be different in the two cases.

Measurements of the change of temperature of a solid on being stretched or compressed were first made by Joule in the 19th century and extensive studies on rubbers by Müller about 20 years ago. For plastics little work has been done until the recent studies of Haward and his associates[40] and the work, using the Tian-Calvet microcalorimeter, of Godovskii et al.[41]

In all these studies uniform stress and uniform temperature are assumed in the sample and the tests are of the Kelvin relation connecting stress and temperature.

The contribution of the thermoelastic effect to energy dissipation in solids under transient or cyclic deformation was first studied by Zener[42] and shown to account for mechanical relaxation peaks in some metals.

The general theory of thermoelasticity is well documented in the book by Nowacki[43] which also gives solutions for some physical problems. In particular those for the quasi-static case in which inertia effects can be neglected are derived from the work of Biot. However, there does not appear to be any experimental work on the relation of internal friction to the thermoelastic effect other than that on metals (Zener[42], Nowick and Berry[44]).

In our own recent low temperature studies[45] of ultra high drawn polyethylene, a change in the background level of tan δ in tensile tests when the environmental gas is changed, is attributed to thermoelastic effects resulting from the different heat transfer characteristics of the gases used.

4 Inhomogeneous Materials

In the previous chapters it has been assumed that the elastic properties of the material were not dependent upon position. We must now consider materials in which this is not true. Examples of polymeric solids in which the elastic constants depend upon position are semi-crystalline polymers, copolymers and composites. In the first of these the elastic properties of the crystalline regions are in general, anisotropic and different from those of the amorphous regions lying between them. If, therefore, a function of position $C_{ijkl}(\mathbf{r})$ is considered, where C_{ijkl} is the generalized coefficient of stiffness (its inverse, the compliance s_{ijkl} may be chosen as well) then in the crystalline regions there will be the usual 81 components of C_{ijkl}, reducing to 21 or less with crystal symmetry, whereas in the amorphous regions the C_{ijkl} will differ from the crystal values and the symmetry will be such as to require only two components. C_{ijkl} (**r**) is therefore a function of position which describes the inhomogeneity of the material.

In copolymers such as styrene-butadiene-styrene the nature of the phase separation is such that c_{ijkl} (**r**) is a periodic function of **r** with the symmetry of the phase lattice – that is, the long-range translational order of the phases.

An example of such order is shown by the hexagonal symmetry of SBS as revealed by LAXD, electron microscopy and mechanical measurements. In composite materials the choice of phase is at the disposal of the material "designer" and the phase lattice and phase geometry may be chosen to optimise desired properties of the material. The reinforcing phase is usually regarded elastically as an "inclusion" in a matrix of the material to be reinforced. In most cases the inclusions do not occupy exactly periodic positions in the host phase so that quasi-hexagonal or quasi-cubic structure is obtained rather than, as in the copolymers, a nearly perfect ordered structure.

From what has been stated above it is clear that there is a similarity in concepts over a considerable range of scale as between, on the one hand "real" crystals where the elements are atoms, through ordered phase materials such as copolymers, where the elements are amorphous but in a "crystalline" lattice, to macro-crystalline materials such as composites. Since a great deal of work has been done in recent years on the properties of composites, in particular fibre composites, it is worth examing the generality of the results obtained particularly in the respect of molecular "composites" such as semi-crystalline polymers and copolymers.

Inhomogeneous materials, both natural and man-made, may be classified according to the following factors:

a) Scale: – macroscopic, microscopic, molecular etc.
b) The spatial distribution of the inhomogeneities: – random, pseudo-crystalline, crystalline.
c) Local geometry: – whether the inhomogeneities are spherical, ellipsoidal, lenticular or needle-shaped and so on.
d) The elastic symmetry of the inhomogeneities: – which may not coincide either with the lattice symmetry, if any, of b) or with the local geometrical symmetry suggested in c).

4.1 Scale

We need to consider this question of scale more broadly when we wish to apply elasticity (or viscoelasticity) theory to real materials. Consider the following solids:
1) Window glass, 2) A polycrystalline metal, 3) Sandstone, 4) Concrete, 5) A brick wall, 6) The surface crust of the Earth, with folds and intrusions.

In each case we can choose a scale above which the material may be considered *homogeneous*, but below which the structure needs to be considered. Roughly, for the list given the scale is: a few nanometres, a few microns, a few millimetres, tens of millimetres, a tenth of a metre, several kilometres.

We never in practice achieve the homogeneity of physical properties we should like to assume in the applications of elasticity theory. Thus the solutions, available in the treatises, of such problems as simple extension, torsion, bending, the effect of point loads, pressurised cylinders and spheres and so on are never exactly applicable to real materials (even to glass) and therefore never verifiable. To quote Love "... many materials used in engineering structures ... do not obey Hooke's law for any strains that are large enough to be observed. It is known also that those materials which do obey the law for small measurable strains do not obey it for larger ones". This is partly the result of scale effects, which we now need to define more exactly. One of the earliest considerations of scale in solids was by Sander[46)] in his work *Gefügekunde der Gesteine*, translatable as structural petrology. It can be said: "a body is statistically homogeneous when the average of the internal constitution in any volume element is the same for all volume elements with dimensions not smaller than the scale under consideration". He uses the term "fabric" (Ger. Gefüge) to denote the internal ordering of physical and spatial data in an aggregate. (The word "texture" is more common in metallurgy and in polymer science).

The components may be crystal grains and their arrangement is not generally subject to the three dimensional periodicity and the energy dependent packing laws of the crystal lattice – so that the permissible kinds of order differ from those of crystals.

The symmetry of a fabric or texture is defined statistically by Sander and expressed in terms of point group symmetry because translational symmetry is absent in petrology.

There are five types:

a) Spherical $K_{\infty h}$ – random orientation of fabric elements.
b) Axial $D_{\infty h}$ – spheroidal symmetry with one unique axis.
c) Orthorhombic D_{2h} – the symmetry of a triaxial ellipsoid.
d) Monoclinic C_{2h} – a single plane of symmetry with a diad axis normal to it.
e) Triclinic $S_2(C_i)$ – no planes of symmetry.

Similar considerations of symmetry apply in other systems, for example nematic liquid crystals and aligned short fibre composites have symmetry $D_{\infty h}$, smectic A liquid crystals D_{∞}, while in copolymers and certain fibre composites examples of hexagonal symmetry may be found and translational symmetry may also be present, which is not found in petrology.

In 1963 Hill[47)] defined the Representative Volume Element (RVE) in a consideration of general properties of composite materials. The definition is more exact than Sander's, which it includes.

The representative volume element is a sample that a) is structurally entirely typical of the whole mixture on average and, b) contains a sufficient number of inclusions for the apparent overall moduli to be effectively independent of the surface values of traction and displacement so long as these values are "macroscopically uniform". That is, they fluctuate about a mean with a wavelength small compared with the dimensions of the sample, and the effects of these fluctations become insignificant within a few wavelengths of the surface. By average over a volume Hill means the integral taken over the volume, divided by that volume.

Thus

$$\langle \sigma_{ij} \rangle = \frac{1}{V} \int_V \sigma_{ij} d\tau$$

The determination of "sufficient number of inclusions" immediately involves questions of scale and also of the method of testing when experiments are being designed, for clearly if ultrasonic methods were used, for example, on a material with phases of a millimetre or so in dimensions then anomalous effects would be observed when frequencies of a megahertz were used. The surface tractions at the phase boundaries would not be macroscopically uniform.

We now consider some models of polymer structure and ascertain their usefulness as representative volume elements. The Takayanagi[48)] series and parallel models are widely used as descriptive devices for viscoelastic behaviour but it is not correct to use them as RVE's for the following reasons. First, they assume homogeneous stress and displacement throughout each phase. Second, they are one-dimensional only, which means that the "modulus" derived from them depends upon the directions of the surface tractions. If we want to make up models such as the Takayanagi ones in three dimensions then we shall have a composite "brick wall" with two or more elements in each of which the stress is non-uniform.

A conscious choice of such elements can be made but in general the equilibrium distribution of stress cannot be found except for particular geometries. The assumptions of uniform strain throughout the assembly or of uniform stress were respectively made by Voigt and by Reuss. Returning to the structures actually perceivable in polymers one may consider the spherulite in a semi crystalline polymer as being unsuitable as a RVE because the boundary is not included. However, an assembly of spherulites would be acceptable, since it would contain sufficient to make it entirely typical of the bulk and because such an assembly would have moduli independent of the surface tractions and displacements. The linear size of such a representative volume element of spherulites would be perhaps several hundred microns.

Intuitively, however, the RVE for an amorphous polymer may be expected to be no more than 10 nanometres or so. In an aligned fibre composite with continuous fibres we need to take a sample sufficiently large as to contain perhaps 30 to 50 fibres in a lateral dimension and in this case the coefficient of variation of stress as between fibres would enable a measure to be decided. In a random array of fibres however (chopped strand mat) the RVE must be of the order of a fibre length at least as is also the case for an aligned short fibre composite.

Where a melt-crystallized polymer has been processed by drawing, rolling or other means to produce an aligned structure in which lamellae as well as polymer chains have discernible order, a pseudocrystalline unit cell is present. Provided that this unit cell contains elements of the crystals as well as the boundaries between crystals and that it is entirely typical of the material as a whole then it could be considered as a RVE within the meaning defined above. The lamella crystal itself sometimes considered as embedded in an amorphous matrix would not seem to be an acceptable RVE for reasons similar to those advanced against the Takayanagi model, namely that its modulus is dependent upon the surface tractions. The boundaries between lamella crystals in the matrix must be included in an acceptable RVE.

Our reason for stressing the concept of representative volume element is that it seems to provide a valuable dividing boundary between continuum theories and molecular or microscopic theories. For scales larger than the RVE we can use continuum mechanics (classical and large strain elasticity, linear and non-linear viscoelasticity) and derive from experiment useful and reproducible properties of the material as a whole and of the RVE in particular. Below the scale of the RVE we must consider the micromechanics if we can – which may still be analysable by continuum theories but which eventually must be studied by the consideration of the forces and displacements of polymer chains and their interactions.

The properties of the RVE may then be deducible from below, as it were, in terms of these molecular quantities.

4.2. Ordered Heterogeneous Materials with Phases Considered as Continua. The Unit Cell Approach

In two dimensions the problem of an array of elements of one material embedded in another has attracted attention because of its relevance to fibre reinforcement. Square arrays and hexagonal arrays of circular or elliptical elements have been analysed using classical elasticity using the complex variable method[49)] and by numerical methods[50)]. The results are also applicable to three dimensional ordered arrays such as are found in filament wound structures and in other situations such as copolymers of styrene and butadiene. Agreement with experiment is generally good except for the transverse properties which are sensitive to the choice of array and the method of calculation.

Analyses of short fibre composites in two dimensions have also been made by various means, usually numerical. The overlapping array is commonly chosen, illustrated in Fig. 9.

A study of the lamella stack in two dimensions was made by one of the authors and Lock[51)]. In this model rectangular elements of two different materials are stacked along the axis of loading and the overall modulus calculated. For completeness the other moduli should be calculated but this was not done.

In three dimensions a treatment, approaching exactness of the short fibre array was made by Smith and Spencer[52)] in which hexagonal packing was assumed but the hexagonal unit cell replaced by a composite circular cylinder. Numerical analysis of the cubic lattice of spheres was made by Broutman and Agarwal[53)].

In most of these unit cell analyses the assumptions are made that a) strains are small, so that infinitesimal elasticity theory applies, b) the phases are of uniform density and elastic properties and c) the phases are isotropic. Some extensions to the cases of anisotropic phases have been made[54)].

The unit cell approach has the following advantages: prediction of the overall elastic moduli in terms of the phase moduli; knowledge of the stress distributions within phases (relevant to the prediction of failure mechanisms); usable formulae in which can be substituted either the phase moduli and concentration or, by working backwards from the overall moduli, from which they can be deduced. From this point of view analytic expressions such as those of Van Fo Fy and Savin are of more use than numerical data (though presumably these could be fitted for comparison purposes by polynomials or other means).

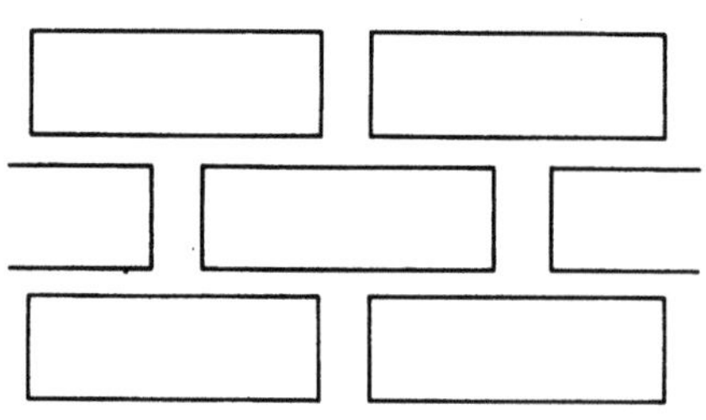

Fig. 9. Overlapping array used in analyses of short fibre composites

The elastic constants derived by Van Fo Fy and Savin are as follows. (The symmetry axis is 3, c is the concentration of the circular reinforcing phase in a hexagonal array. The compliance constants S_{ij} are quoted)

$$
\begin{aligned}
S_{11} = S_{22} &= \frac{(1-\nu_2)^2}{E_2}\left[\frac{2+(\alpha_1-1)G_2/G_1}{2-c+c\alpha_2+(1-c)(\alpha_1-1)G_2/G_1} - \frac{2c(1-G_2/G_1)}{c+\alpha_2+(1-c)G_2/G_1}\right] \\
S_{44} = S_{55} &= \frac{1-c+(1+c)G_2/G_1}{1+c+(1-c)G_2/G_1}\cdot\frac{1}{G_2} \\
S_{66} &= \frac{(1-c)\alpha_2+(1+c\alpha_2)G_2/G_1}{c+\alpha_2+(1-c)G_2/G_1}\cdot\frac{1}{G_2} \\
S_{23} = S_{32} \sim S_{13} = S_{31} &= -S_{22}\left[\nu - \frac{4c(1-\nu_2)(\nu_2-\nu_1)}{2-c+c\alpha_2+(1-c)(\alpha_1-1)G_2/G_1}\right] \\
S_{33}^{-1} &= cE_1 + (1-c)E_2 + \frac{8c(1-c)G_2(\nu_1-\nu_2)^2}{2-c+c\alpha_2+(1-c)(\alpha_1-1)G_2/G_1} \\
S_{12} = S_{21} &= S_{33}\left[\nu_2 - \frac{c(1+\alpha_2)(\nu_2-\nu_1)}{2-c+c\alpha_2+(1-c)(\alpha_1-1)G_2/G_1}\right]^2 \\
&\quad - \left(\frac{1+\alpha_2}{E_2}\right)\left\{\nu_2 + (1-\nu_2)c\left[\frac{\alpha_2-1-(\alpha_1-1)G_2/G_1}{2-c+c\alpha_2+(1-c)(\alpha_1-1)G_2/G_1} - \frac{2(1-G_2G_1)}{c+\alpha_2+(1-c)G_2/G_1}\right]\right\}
\end{aligned}
\tag{4.1}
$$

In these expressions E_i, G_i, ν_i are respectively Young's and shear moduli and Poisson's ratio for phase i, $\alpha_i = 3\text{–}4\,\nu_i$ and

$$
\begin{aligned}
S_{12} &= -\nu_{12}S_{11} = S_{21} = -\nu_{21}S_{22} \\
S_{23} &= -\nu_{23}S_{22} = S_{32} = -\nu_{32}S_{33} \\
S_{31} &= -\nu_{31}S_{33} - S_{13} = -\nu_{13}S_{11}
\end{aligned}
$$

The explicit formulae given by Rosen[55] are also of value. They are derived from a model consisting of a random assemblage of composite cylinders (Hashin and Rosen[56]) and expressed in terms of the axial Young modulus E_a^*, the Poisson ratio for uniaxial stress in the fibre direction ν_a^*, the transverse plane strain bulk modulus k_t^*, the axial shear modulus G_a^* and the transverse shear modulus G_t^*.

$$
\begin{aligned}
E_a^* &= \overline{E} + \frac{4c(1-c)(\nu_f-\nu_m)^2}{(1-c)/k_f + c/k_m + 1/G_m} \\
\nu_a^* &= \bar{\nu} + \frac{c(1-c)(\nu_f-\nu_m)(1/k_m - 1/k_f)}{(1-c)/k_f + ck_m + 1/G_m} \\
k_t^* &= \frac{k_m k_f + \overline{k}G_m}{(1-c)/k_f + c/k_m + G_m}
\end{aligned}
\tag{4.2}
$$

$$G_a^* = G_m \frac{(1-c)G_m + (1+c)G_f}{(1+c)\,G_m + (1-c)G_f}$$

$$G_t^* = G_m \frac{(\alpha + \beta_m c)(1 + \varrho c^3) - 3c(1-c)^2\beta_m^2}{(\alpha - c)(1 + \varrho c^3) - 3c\,(1-c)^2\beta_m^2}$$

where

$$\alpha = \frac{\gamma + \beta_m}{\gamma - 1}, \qquad \beta = \frac{1}{3 - 4\nu}$$

$$\varrho = \frac{\beta_m - \gamma\beta_f}{1 + \gamma\beta_f}, \qquad \gamma = G_f/G_m$$

c = concentration of fibres.

The overbar in quantities such as $\overline{E}$, $\bar{\nu}$ denotes the volume average of the quantity. Thus $\overline{E} = c\,E_f + (1-c)E_m$. The subscripts f, m denote fibres and matrix respectively. Derived from the above is the transverse Young modulus

$$E_t^* = \frac{4k_t^* G_t^*}{k_t^* + G_t^*(1 + 4k_t^* \nu_a^{*2}/E_a^*)} \tag{4.3}$$

and the Poisson ratio ν_t^* in the transverse plane

$$\nu_t^* = \frac{1}{2}\,(E_t^*/G_t^* - 1) \tag{4.4}$$

These formulae and some others were compared with measurements of the elastic moduli of a SBS copolymer showing hexagonal symmetry[12)].

4.3 Bounds on Elastic Constants

Where exact solutions cannot be found either because no lattice symmetry is present or because, even if there is a space lattice, there is no point group symmetry, or if both are absent, there are certain bounds on elastic constants which have been found. A great deal of progress has been made since the early part of the century when the uniform strain (Voigt) and uniform stress (Reuss) bounds were postulated. Improved bounds have been found by Hill[47)], Hashin and Shtrikman[57)], Walpole[58)], Kröner[59)] and others. In some cases these bounds also apply (Roscoe[60)]) to the complex moduli. The bounds are obtained by the application of theories from classical infinitesimal elasticity theory – the principles of minimum potential energy and of minimum complementary energy. The most elementary application of these principles provides the familiar Voigt and Reuss bounds. For the Voigt bound for example, the strains in each phase are assumed uniform and equal to the mean overall strain. The principle of minimum potential energy states that for the *equilibrium* state the strain energy must be less than that for any state in which the equations of equilibrium are not satisfied. Such a state is that of uniform strain in all the phases. Since then the stresses will not be in equilibrium between phases and any other state must have lower energy, so that the Voigt assumption leads to an upper bound on the elastic modulus. Similarly the Reuss estimate (uniform stress in all phases) leads to a lower bound. However, by considering states of strain (still uniform in each phase) other than those above it is possible to arrive at better bounds. This was done first by Hashin and Shtrikman, Hill and generalized by Walpole[61].

For isotropic materials the two moduli to be considered are the bulk modulus K and the shear modulus G. Bounds are proved by Walpole[61] as follows:

$$\frac{c_1}{1 + (K_1 - K_2)c_2/(K_2 + K_\ell^*)} \leqslant \frac{K - K_2}{K_1 - K_2} \leqslant \frac{c_1}{1 + (K_1 - K_2)c_2/(K_2 + K_g^*)}$$

$$\frac{c_1}{1 + (G_1 - G_2)c_2/(G_2 + G_\ell^*)} \leqslant \frac{G - G_2}{G_1 - G_2} \leqslant \frac{c_1}{1 + (G_1 - G_2)c_2/(G_2 + G_\ell^*)} \tag{4.5}$$

In these expressions c_i, K_i, G_i are, respectively, the concentration, bulk modulus and shear modulus of phase i.

$K_g^* = 4/3\, G_g$ and $K_\ell^* = 4/3\, G_\ell$

and

$$G_\ell^* = \frac{3}{2}\left(\frac{1}{G_\ell} + \frac{10}{9K_\ell + 8G_\ell}\right)^{-1}$$

$$G_g^* = \frac{3}{2}\left(\frac{1}{G_g} + \frac{10}{9K_g + 8G_g}\right)^{-1}$$

G_g, K_g are the greatest of G_i, K_i } in the mixture
G_ℓ, K_ℓ are the least of G_i, K_i }

These results of Walpole[61] include as special cases those of Hill[47] and of Hashin and Shtrikman[48]. For anisotropic phases Walpole[58] gives bounds on the five elastic moduli of an aligned array of transversely isotropic elements and for randomly oriented fibrous inclusions in an isotropic matrix. For the former case (alignment) the bounds are expressed in terms of phase concentration c_i and the quantities k, l, m, n, p defined as follows: $k = 1/2\,(C_{11} + C_{12})$, $m = 1/2\,(C_{11} - C_{12})$, $\ell = C_{13}$, $n = C_{33}$, $p = C_{44} = C_{55}$.

Because of the asumed transverse isotropy it follows that $C_{66} = 1/2\,(C_{11} - C_{12})$. The terms C_{ij} are the elastic stiffnesses expressed in the contracted (Voigt) notation.

(i.e. $C_{11} = C_{1111}$, $C_{12} = C_{1122}$, $C_{44} = C_{2323}$ etc.)

Then the bounds proved by Walpole are

$$\bar{k} = [\Sigma c_i/(k_i + m_o)]^{-1} - m_o$$

$$\bar{l} = [\Sigma c_i l_i/(k_i + m_o)]/[\Sigma c_i/(k_i + m_o)]$$

$$\bar{n} = \Sigma c_i n_i + \frac{\{\Sigma c_i n_i[\Sigma c_s(l_s - l_i)/(k_s + m_o)]/(k_i + m_o)\}}{[\Sigma c_i/(k_i + m_o)]}$$

$$\bar{m} = \{\Sigma c_i/[m_i + m_o k_o/(k_o + 2m_o)]\}^{-1} - m_o k_o/(k_o + 2m_o)$$

$$\bar{p} = [\Sigma c_i/(p_i + p_o)]^{-1} - p_o \tag{4.6}$$

Here the upper bound is attained where p_o, ℓ_o, m_o are the *greatest* of the values p_i, ℓ_i, m_i, and the lower bound is attained where p_o, ℓ_o, m_o are the *least* of these. Note, they do not need to belong to the same phase.

Walpole also gives bounds for randomly oriented fibrous inclusions in an isotropic matrix but these are not easily stated and the reader is referred to the original paper (Walpole[58]).

Roscoe[60] and Laws and McLaughlin[30] have considered the problem of linearly viscoelastic elements, where the extremum theorems of elasticity theory do not apply. Roscoe considers linear viscoelasticity and uses the complex modulus comparing the material with an elastic one of the same phase geometry. He shows that the real parts of the overall moduli of the viscoelastic composite are not less than the corresponding overall moduli of the elastic composite when its phases have moduli equal to the real parts of the moduli of the corresponding phases in the viscoelastic composite. Similarly for the imaginary parts.

Results of corresponding form can be derived for compliances instead of moduli. Thus if the complex moduli of the composite are

$$G^* = G' + iG'' , \qquad K^* = K' + iK''$$

we have

$$\begin{aligned} G' &\geq G_\ell(G_r', K_r') , \qquad G'' \geq G_\ell(G_r'', K_r'') \\ K' &\geq K_\ell(G_r', K_r') , \qquad K'' \geq K_\ell(G_r'', K_r'') \end{aligned} \tag{4.7}$$

where $G_\ell(G_r', K_r')$ represents a lower bound on the overall rigidity of an *elastic* composite with phases having rigidities G_r' and bulk moduli K_r' etc. Similarly upper bounds are derived from the corresponding relations for compliances. The bounds (upper and lower) used in the treatment may be the elementary Voigt and Reuss bounds or the improved bounds of Hashin and Shtrikman, Hill and Walpole discussed above.

Laws and McLaughlin[30] discuss viscoelastic creep compliances of composite materials using another approach to the problem of the elastic properties of heterogeneous materials – the self-consistent method.

4.4 Self-Consistent Method

This method relies on the exact solution of the elastic problem for an inclusion of known geometry (an ellipsoid) surrounded by an infinite matrix. The composite problem to be solved is that in which the included phases are ellipsoidal in shape. Selecting one as the reference ellipsoid, the effect of the remainder is approximated by a continuum surrounding the reference ellipsoid, thus reducing the problem to one for which there is an exact solution. Finally, the result obtained has to be self-consistent with the properties assumed for the continuum.

The method was used first by Kröner[59)] for the elastic moduli of cubic polycrystals and has been applied to composite materials by Hill[62)], Walpole[58)] and many others (see e.g. Laws and McLaughlin[30)] for references). An outline of the theory derived mainly from Walpole's paper is given in § 4.5 below. Here we quote Walpole's formulae for an arbitrary dispersion of spherical inclusions dispersed throughout a matrix of another material in a homogeneous and isotropic distribution on average. The bulk modulus K and shear modulus G are given as

$$K = K_2 + \frac{c(K_1 - K_2)}{1 + (K_1 - K_2)(1 - c)/(K_2 + K^*)}$$
$$G = G_2 + \frac{c(G_1 - G_2)}{1 + (G_1 - G_2)(1 - c)/(G_2 + G^*)} \qquad (4.8)$$

where

$$K^* = 4/3\,G, \qquad G^* = 3/2(1/G + 10/(9K + 8G))^{-1}$$

subscripts 1 and 2 rcfcr to spheres and matrix respectively, c is the concentration of spheres.

The equations are not, of course, explicit since each contains a starred term which is a function of both K and G.

For aligned transversely isotropic elements the self consistent method gives (Walpole[58)]) the relations

$$\frac{c}{p - p_2} + \frac{1 - c}{p - p_1} = \frac{1}{2p}$$
$$2\left[\frac{c}{m - m_2} + \frac{1 - c}{m - m_1}\right] = \frac{1}{k + m} + \frac{1}{m}$$
$$\frac{c}{k_1 + m} + \frac{1 - c}{k_2 + m} = \frac{1}{k + m} \qquad (4.9)$$
$$E - c\,E_1 - (1 - c)E_2 = \frac{4c(1 - c)(\nu_1 - \nu_2)^2}{c/k_2 + (1 - c)/k_1 + 1/m}$$
$$\nu - c\nu_1 - (1 - c)\nu_2 = \frac{c(1 - c)(\nu_1 - \nu_2)(1/k_2 - 1/k_1)}{c/k_2 + (1 - c)/k_1 + 1/m}$$

where the quantities c, p, m, k are as defined above (Eq. 4.6) and ν, E are Poisson's ratios and Young's moduli respectively.

Again the equations are not in explicit form.

Laws and McLaughlin[30)] solve the problem of the viscoelastic ellipsoidal inclusion in anisotropic materials and then use the self consistent method to calculate the overall viscoelastic compliances for a composite.

4.5 The Self-Consistent Theory Involves Three Stages

a) The misfitting inclusion in an infinite matrix.

b) The perfect inclusion under strain at infinity.

c) The "self-consistent" problem of an inclusion in a matrix of elastic properties equal to the composite of similar inclusions.

a) Consider an inclusion of arbitrary shape in an infinite matrix. Let its elastic modulus tensor be denoted L_1 with inverse (compliance) tensor M_1 and let the matrix have elastic tensors L and M.

Now suppose the inclusion to be removed from its environment, deformed by a strain e and then replaced. The stress s required to return the deformed inclusion to its original shape will be given by $-L_1e$ and the equilibirum state of the inclusion after replacement will be with strain ε and stress $\sigma = L_1\varepsilon - L_1e$.
The average strain $\bar{\varepsilon}$ and average stress $\bar{\sigma}$ in the inclusion are then defined as

$$\bar{\varepsilon} = P_1 \, S = P_1 \, L_1 \, e$$

$$\bar{\sigma} = -Q_1 \, e = -QM_1 \, s$$

Then

$$\begin{aligned} \bar{\sigma} &= L_1 \, \bar{\varepsilon} - L_1 \, e = (L_1P_1L_1 - L_1)e \\ &= -Q_1e \end{aligned}$$

Hence

$$\begin{aligned} Q_1 &= L_1 - L_1P_1L_1 \\ M_1Q_1 &= I - P_1L_1 \end{aligned}$$

The sequence so far is as illustrated

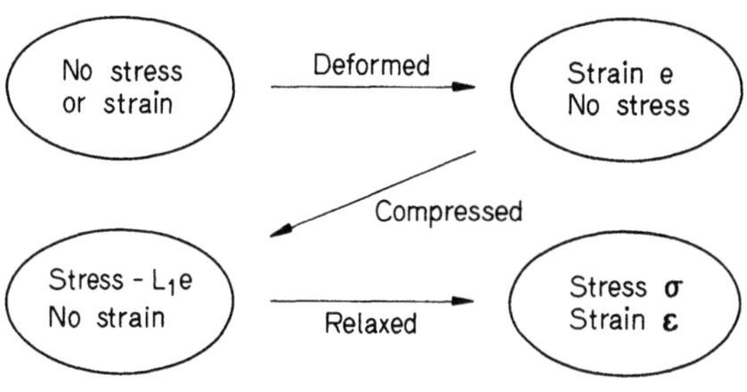

b) Now consider a perfectly fitting inclusion and apply at infinite distance a uniform strain field ε^A with stress $\sigma^A = L\varepsilon^A$ to the matrix in which the inclusion is held.

The inclusion has now become a *misfitting* inclusion in a medium of elastic constants L with strain ε^A, stress σ^A. The stress s used in the previous part now becomes

$$\sigma^A = L_1\varepsilon^A$$

and the average strain in the inclusion becomes

$$\begin{aligned}\bar{\varepsilon}_1 = \bar{\varepsilon} + \varepsilon^A &= P_1(\sigma^A - L_1\varepsilon^A) + \varepsilon^A \\ &= (P_1(L - L_1) + I)\varepsilon^A \\ &= A_1\varepsilon^A\end{aligned}$$

and the stress $\bar{\sigma}_1 = \bar{\sigma} + \sigma^A$

$$\begin{aligned}&= (Q_1(M - M_1) + I)\sigma^A \\ &= B_1\sigma^A\end{aligned}$$

Then $L_1\bar{\varepsilon}_1 = \bar{\sigma}_1 = L_1A_1\varepsilon^A = B_1L\varepsilon^A$

So that $L_1A_1 = B_1L$

and similarly, $M_1B_1 = A_1M$

In principle the tensors P_1 and Q_1 can be calculated for any shape but spheres and ellipsoids are the usual shapes used with cylinders and discs as the limiting cases. Walpole[58)] gives calculations for the latter, Eshelby[63)] for spheres, Kröner[59)] for spheroids.

c) The self-consistent problem. Consider two phases, the inclusion and the matrix, with constants L_1, L_2, (M_1, M_2) respectively. Let the concentration of phase 1 be c_1 and that of 2, c_2.

The average strain in the inclusion is given by $\bar{\varepsilon}_1 = \{\overline{A}_1\}\ \bar{\varepsilon}$ and in the matrix by $\bar{\varepsilon}_2 = \overline{A}_2\ \bar{\varepsilon}$, where $\bar{\varepsilon}$ is the average overall strain and { } denotes the average over all orientations and inclusion shapes. (The corresponding stresses are $\{\overline{B}_1\}\ \bar{\sigma}$ and $\overline{B}_2\bar{\sigma}$ respectively) Then

$$c_1\{\overline{A}_1\} + c_2\overline{A}_2 = I = c_1\{\overline{B}_1\} + c_2\overline{B}_2$$

This may be seen as follows.

The overall average strain $\bar{\varepsilon} = \frac{1}{V}\int \varepsilon d\tau$

Hence $\bar{\varepsilon} = \frac{1}{V}\int\sum_i \varepsilon_i d\tau$

which by summing over the phases

$$= \frac{1}{V}\sum_i\int \varepsilon_i d\tau = \frac{1}{V}\sum_i V_i\bar{\varepsilon}_i = \sum_i c_i\bar{\varepsilon}_i = \sum c_i\{\overline{A}_i\}\bar{\varepsilon}$$

or $\sum c_i\{\overline{A}_i\} = I$ Similarly for $\{\overline{B}_i\}$

We also have the average stress

$$L\bar{\varepsilon} = \bar{\sigma} = \frac{1}{V}\sum\int L_i\varepsilon_i d\tau = \frac{1}{V}\sum L_i\int \varepsilon_i d\tau$$

$$= \frac{1}{V}\sum L_i V_i \bar{\varepsilon}_i = \sum c_i\{L_i\overline{A}_i\}\bar{\varepsilon}$$

so that

$$L = \sum c_i\{L_i\overline{A}_i\}$$

Therefore in the present case

$$L = c_1\{L_1\overline{A}_1\} + c_2L_2\overline{A}_2$$

and

$$M = c_1\{M_1\overline{B}_1\} + c_2M_2\overline{B}_2$$

Now the self consistent approach assumes that $\overline{A}_1 = A_1$, $\overline{B}_1 = B_1$

Eliminating $\overline{A}_2$, $\overline{B}_2$ we have

$$L = c_1\{L_1\overline{A}_1\} + L_2(I - c_1\{\overline{A}_1\})$$
$$M = c_1\{M_1\overline{B}_1\} + M_2(I - c_1\{\overline{B}_1\})$$

and therefore

$$L - L_2 = c_1\{(L_1 - L_2)\overline{A}_1\}$$
$$M - M_2 = c_2\{(M_1 - M_2)\overline{B}_1\}$$

Now

$$\overline{A}_1 = A_1 = P_1(L - L_1) + I$$
$$\overline{B}_1 = B_1 = Q_1(M - M_1) + I$$

Hence

$$L - L_2 = c_1(L_1 - L_2)\{[P_1(L - L_1) + I]\}$$
$$M - M_2 = c_1(M_1 - M_2)\{[Q_1(M - M_1) + I]\}$$

or

$$[I - c_1\{(L_1 - L_2)P_1\}]L = c_1L_1 + (1 - c_1)L_2 - c_1\{(L_1 - L_2)P_1L_1\}$$
$$= L_v - c\{(L_1 - L_2)P_1L_1\}$$

and

$$[I - c_1\{(M_1 - M_2)Q_1\}]M = M_R - c_1\{(M_1 - M_2)Q_1M\}$$

where L_V and M_R are the Voigt and Reuss sums of the L, M respectively.

Now M is L^{-1} so that the two equations may be combined to give self consistent expressions for L in terms of the L_i or M in terms of the M_i

$$L = L_V - c_1[I - c_1\{(L_1 - L_2)P_1\}]^{-1}\{(L_1 - L_2)P_1(L_1 - L_V)\}$$

$$M = M_R - c_1[I - c_1\{(M_1 - M_2)Q_1\}]^{-1}\{(M_1 - M_2)Q_1(M_1 - M_V)\}$$

5 The Representative Volume Element for Solid Polymers

We need to consider the size of the representative volume element (RVE) for polymers bearing in mind the requirements of its definition. These were 1) the RVE should be entirely typical of the material on average; 2) it must contain sufficient of the inclusions (phases) for the overall moduli to be independent of the surface tractions and displacements provided these are macroscopically uniform.

Amorphous polymers, as the name implies, are structureless except at the molecular level where we shall propose a suitable RVE. Semicrystalline polymers exhibit a wide variety of structures depending upon their chemical nature, the degree of polymerization, the form and size of crystals and their assembly into spherulites, lamellae, fibrils etc.

We may classify the structures in solid polymers as follows in an attempt to include all possibilities.

Structure	*Suggested linear dimension of representative volume element.*
Amorphous homogeneous	
1. Structureless	Half the RMS end-to-end distance of chains
2. clusters present	The mean cluster separation or a small multiple of it
3. cross linked	The network mean mesh size.
Amorphous heterogeneous	
1. Copolymers without phase separation	Half the RMS end-to-end distance of the copolymer chain.
2. Copolymers with phase separation	The mean distance between phases
3. Composites, including foams.	As above interpreting the separate components as phases.
Semi-crystalline random	
1. fringed micelle structure	Sufficient micelles, each about 1000 Å for homogeneity say 5000–10,000 å
2. gels	The equivalent mesh size of the gel
3. lamellar phases	Sufficient lamellae to ensure homogeneity, say 1μ
4. dendritic phases	Several dendrites of about 100μ, hence $500–1000\mu$
5. spherulites	Similar to above
6. Liquid crystal structures	Similar to the scale for lamellae, probably 1μ.

Semi-crystalline oriented

1. broken lamellae	Probably similar to the scale for micelle structures.
2. fibrous, part oriented,	Of the order of the mean fibre length
3. fibrous, fully oriented	Sufficient to include the substructure of fibres
4. row crystallization	The row width or a multiple thereof.
5. shish-kebab structures	If randomly oriented, the RMS end to end distance of the fibrils.

Highly perfected crystalline

1. Solution grown lamellae	The lamella size – 100–500 Å
2. pressure crystallized	A multiple of the lamella thickness, say 1 μ
3. Solution grown fibrillar	The crystal unit cell.

In practice the structure of any given polymer sample is by no means as regular as the above classification would imply and in most cases defies description in terms of recognizable structural elements. For example, Wunderlich[64)] shows examples of "cobweb" structures which can be found in polymers. Clearly, for the purposes of research specific structures have been identified and studied – but this does not mean that a regular solid of macroscopic dimensions may be contructed with these structures. In polymers therefore, we always have to deal with statistical assemblies of elements more or less precisely defined as e.g. lamellar crystal, fibrous crystals, tie chains etc.

To define a representative volume element of such a material is therefore not easy if the aim is to be able to calculate its properties from a knowledge of its structure and the properties of its elements. A number of attempts to do this have nevertheless been made and some success at explaining the physical properties of polymers achieved thereby.

5.1 Amorphous Polymers

Arridge[65)] argues for taking the linear dimension of the RVE for an amorphous polymer as $1/2\,\langle r^2\rangle^{1/2}$, where $\langle r^2\rangle$ is the mean square end-to-end distance of the polymer chains, for the following reason. If we consider a cube of side a in an amorphous polymer and calculate the probability p that a polymer chain entering one face of the cube leaves by the opposite face then we find, to a good approximation,

$$p = (1 - R)R^2$$

where R is given (Haward, Daniels and Treloar[66)])
by the expression

$$R = 1 - 8 \sum_{p=0}^{\infty} (-1)^{r-1} \frac{A(rb)}{b + 2\sqrt{2/\pi}}$$

where $\quad b = \sqrt{3}\,a/(\ell\sqrt{n})$

$$A(u) = 1/\sqrt{2\pi}\,\exp\,(-u^2/2) - u\,[1 - \Phi(u)]$$

and $\quad \Phi(u) = 1/\sqrt{2\pi} \int_{-\infty}^{u} \exp(-t^2/2)dt$

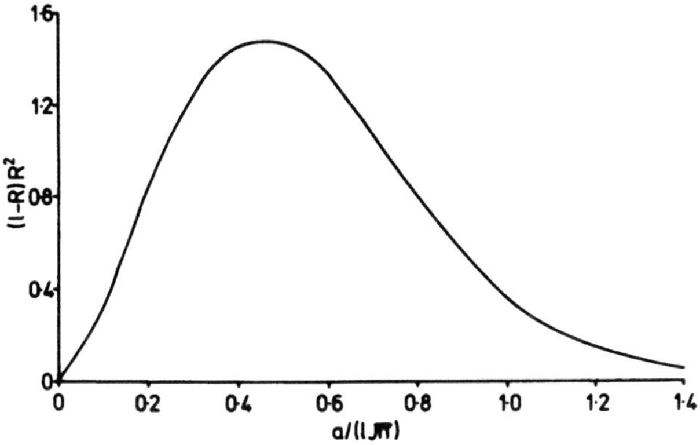

Fig. 10. Probability of chain of n links each of length ℓ crossing a cube of side a as function of $a/(\ell\sqrt{n})$

The dependence of p on the parameter $a/(\ell\sqrt{n})$ is shown in Fig. 10, from which it is clear that it has a maximum for $a/\ell\sqrt{n} = 0.5$, for which value $p \sim 1/7$. It seems logical to choose the cube of side $a = 0.5\ \ell\sqrt{n} = 0.5\ \langle r^2\rangle^{1/2}$ as the RVE in an amorphous polymer.

A dimension much smaller than this will not contain sufficient chains which traverse from face to face carrying load, whereas a dimension much larger will contain no chains at all that traverse, and will therefore be atypical. $\langle r^2\rangle^{1/2}$ for Gaussian chains is of order 200–300 Å for typical molecular weights of 250 000 so that representative volume element dimensions of the order of 100–200 Å are to be expected, depending, of course, on molecular weight.

Computer modelling in the field of amorphous semiconductors has already been done using as many as 1350 atoms (Gaskell, Gibson and Howie[67]) and the same type of modelling applied to polyethylene would imply that molecular weights of about 6000 could be used in such studies. This would imply an RVE of about 34 Å only but it is possible that improved techniques could increase this figure towards the size suggested above.

5.2 Amorphous Copolymers

In the case of block copolymers phase separation leads to partly ordered structures of one phase in a matrix of the other (Hendus[68], Matsuo[69]). While it is known that the elastic moduli of copolymers increase as the proportion of the glassy component increases (Dawkins[70], Allport and Mohajer[71]) few calculations of overall moduli in terms of the constituent moduli seem to have been made. Where these phases are regular spheres or cylinders calculations of the overall elastic properties are possible using the theories detailed in Chap. 4. Arridge and Folkes[72] made such a calculation for an extruded SBS copolymer in which a highly developed hexagonal array of cylinders of polystyrene in a matrix of polybutadiene was found. They measured Young's modulus E_θ on samples cut at various angles to the symmetry axis fitting the results by the relation

$$E_\theta^{-1} = S_{11}\sin^4\theta + (2S_{13} + S_{44})\sin^2\theta\cos^2\theta + S_{33}\cos^4\theta$$

From this equation values of S_{33}, S_{11} and $2S_{13} + S_{44}$ were found and, assuming that $S_{13} \ll S_{44}$, a reasonable assumption on physical grounds, the values of S_{33}, S_{44} and S_{11} were then available to compare with fibre composite predictions. An unexpected byproduct of this study was the discovery that S_{44} measured as above differed by a factor of 2 from its value derived from a torsional shear experiment. This discrepancy was later resolved (see Ref. 12) by the realisation that St. Venant's principle normally invoked to account for end effects requires modification in the case of highly anisotropic solids.

5.3 Semi-Crystalline Polymers

In the field of semi-crystalline polymers several workers have used composite theories to explain their elastic properties in terms of those of those of the crystalline and amorphous phases.

Gray and McCrum[73)] used the Hashin-Shtrikman theory to explain the origin of the γ relaxation in PE and PTFE, Maeda et al.[74)] have given exact analyses of several two phase models for semi-crystalline polymers and Buckley[75)] represented a biaxially oriented sheet of linear polyethylene by a two phase composite model.

Andrews[76)] gave results of the work of Reed and Martin on *cis*-polyisoprene specimens crystallized from a strained cross linked melt and on solid state polymerized polyoxymethylene respectively, explaining the results by simple two phase models. He also summarized the studies of Patel and Philips[77)] on spherulitic polyethylene which showed that the Young's modulus increased as a function of crystallite radius by a factor of 3 up to a radius of about 13 μ and then decreased on further increasing spherulite size.

The results of Patel and Philips were well described by an application of the theory of Halpin and Kardos[78)], who earlier had explained spherulite properties using a model in which crystalline filaments were embedded isotropically in an amorphous matrix.

Analyses of particular geometries were made by Owen and Ward[79)] and Arridge[51)] both on lamella stacks. The earlier work of Raumann and Saunders[80)] and Ward[81)] related to orientation effects in polymers and in Ward's work the Voigt and Reuss bounding scheme was used to provide overall elastic averages, taking into account the effects of orientation on an assumed aggregate of elements of known elastic properties (but unknown size). An extension of this approach was made by McCullough et al.[82)] and by Seferis McCullough and Samuels[83)] using orientation distribution functions, the Voigt and Reuss bounds and the Halpin-Kardos[84)] simplification of composites theories.

Ward and his associates[35)] have also used two phase theories to explain the viscoelastic properties of polymers.

In the case of well-developed fibre symmetry several attempts have been made to explain the overall elastic properties in terms of constituent fibrils or other structures. Barham and Arridge[85, 86)] assumed highly drawn polymers (PE, PP) to consist of a stiff needle-like phase embedded in a softer matrix although they were careful not to identify the needle-like phase with any particular polymer structure beyond saying that it was "not incompatible with existing fibril models" for example those of Peterlin[87)]. The theory was successful in explaining a number of the properties of drawn fibres even though it relied upon the simple shear lag (one dimensional) theory originally due to Cox[88)] and used for fibre composites. In view of the unreality of the authors' assumptions as to real structural elements in the polymer more sophisticated analysis was not justified. The treatments of Gibson, Davies and Ward[37)] and of Peterlin[89)] of the elastic properties of ultra oriented polymers differed from those of Arridge and Barham in that they used the Takayanagi models either as a descriptive device or to analyse a supposed crystal structure. In both cases explanation of the observed physical properties was obtained. At present it is still not possible to provide any more exact analysis of the properties of highly oriented polymers until better structural evidence is available. For a recent review see Ciferri and Ward[90].

The foregoing summary of applications of composites theory to polymers does not claim to be complete. There are many instances in the literature of the use of bounds, either the Voigt and Reuss or the Hashin-Shtrikman, of simplified schemes such as the Halpin-Tsai formulation[84], of simple models such as the shear lag or the two phase block and of the well-known Takayanagi models. The points we wish to emphasize are as follows.

1) No model is really exact unless the geometry of the phases is known as well as the elastic properties of the phase *in its form as present* (which may differ from the values in a bulk phase).
2) The size of the phases is an important factor. If they are too small to be considered as elastic continua then composites theory cannot be applied in its usual form.
3) In general only bounds may be applicable, not exact theories, though modifications of the self-consistent scheme could well be used.
4) In most simple applications of models all that is really being achieved is curve fitting. This applies to the Takayanagi models (which are one dimensional and assume uniformity of stress and strain within each element) as well as to simple fibre models such as the shear lag.

Any exact theory, unless the geometry is simple, involves hopelessly complicated calculations of stress distributions even if the elements are large enough for these to be valid (which is not the case for small assemblies of polymer chains). In principle (see e.g. Chen and Young[91]) any geometry may be treated, but ellipsoids and parallelepipeds are the most usual.

6 Calculation of the Elastic Constants of Polymers

From what has been written in the previous chapters it will be clear that the calculation of the elastic constants of polymers in any exact sense is a formidable task except in the case of idealized chains or perfect crystals of infinite extent. Any other geometry must involve assumptions as to the uniformity of stress throughout the assemblage and indeed, the meaning of the term stress at the molecular level. The value of the elastic modulus in the chain direction is, however, of great interest particularly in view of recent experimental studies of highly oriented polymers produced by various techniques[92–98]. These have shown that very stiff fibres of common polymers may be obtained, and these are of commercial significance in many fields. Theoretical estimates of what might be achieved with perfect chain alignment are therefore of more than academic interest.

What is involved in the calculation of modulus? First, we mean by modulus in materials science a relation between stress and strain in a bulk sample under practical conditions. This means in effect a testing rate or frequency usually less than a few kilohertz or at the most in the ultrasonic region of, say, 10 MHz. We also suppose the sample to be a representative volume element of size suitable for the test method and we assume its elastic properties to be uniform over this RVE.

If therefore we take as RVE a *single chain* then it should be considered in isolation (which is impossible in fact since there is always interaction between a chain and its neighbours whether it is in a crystal, in liquid state or in solution). The alternative is to idealize a chain-in-crystal as being a single chain.

Now at present we have no means of applying a known force to a particular chain and measuring its extension. What can be done is to measure the vibrations of chains in various situations and relate these directly via normal mode analysis, or indirectly via continuum (rod) models to a set of stiffnesses.

Now the frequencies at which measured molecular and lattice vibrations occur are vastly different from those at which we wish to measure the practical elastic constants.

Typical infra red vibration frequencies are of order 10^{13} Hz and the so called accordion mode in paraffins is in the range 100 to 20 cm^{-1} or 3×10^{12} to 6×10^{11} Hz so that these vibrations are at least 6 orders of magnitude faster than ultrasonic measurements. Nevertheless, unless there are known processes giving dispersion between the infra red / Raman frequency range and the "practical" frequency range, we may use the spectroscopic data. (Although the intervening frequency range is not accessible to mechanical tests it is accessible to measurements by other techniques, for example microwave and dielectric spectroscopy).

There are two methods which have been used to date. In the first the force constants derived from IR and/or Raman data are substituted into expressions which give *chain* extension in terms of applied force. A chain modulus is then deduced by using an effective chain cross-sectional area. In the second method lattice dynamics is used to derive all the elastic constants for an infinite crystal.

The calculations of Treloar[99)] on polyethylene, nylon 66 and cellulose used bond angles and lengths determined from X-ray diffraction studies of the crystalline forms of these polymers together with force constants for bond stretching and bending derived from infra-red spectroscopy. He treated the chain as a structure of rods, each capable of lengthwise extension but not bending, connected by torsional springs at their junctions. He removed certain errors in previous similar treatments and his estimates were therefore more accurate but, as he pointed out, depended strongly on the published data on force constants. Further, since he neglected the secondary forces between atoms in adjacent chains, his calculation was not valid as a measure of the modulus of the crystal. Treloar found $E = 1.82 \times 10^{11}$ Nm^{-2} for polyethylene using the force constants published by Rasmussen[100)] and a chain mean cross sectional area of 18.24×10^{-16} cm^2.

A much more recent estimate of the "modulus" of a chain was given by McCullough, Eisenstein and Weikart[101)] who considered the forces not only between backbone "skeletal" atoms but also between the attached hydrogen atoms, using force constants derived from the work of Lifson and Warshel[102)] and from Williams[103)]. The rather curious assumption is made, however, that the C–C bond is rigid on the grounds that "the force constants for bond-length deformation are an order of magnitude greater than (those) for bond-angle deformation".

A Young's modulus of 380 GPa at 0 K is found for the chain under these assumptions. Inasmuch as the force constants used in McCullough et al were derived by Lifson and Warshel from normal coordinate analysis of infra-red spectra and in this derivation the C–C bond was *not* assumed to be rigid the use of Lifson and Warshel's force constants in the way in which McCullough et al. use them would seem to be questionable.

The same assumptions are not made in the work of Shimanouchi and his collaborators[104–106] who calculate the chain modulus in two different but related ways. They use as data the Raman longitudinal acoustic vibrations in a range of n-alkanes going from C_4 to C_{94}. In the first series of studies[104] the lowest Raman frequency was shown to be approximately inversely proportional to the number of carbon atoms in the chain and this proportionality becomes more accurate the larger value of n. In fact for $n > 12$[106] it is quite correct to write $\nu = K/n$. This suggests the vibrations correspond to that of a rod of length $n\ell$, where ℓ is the carbon-carbon separation along the chain and the formula for the equivalent Young's modulus E of such a rod is given by $E = \varrho\,(2\,n\ell\nu)^2$, where ϱ is the crystal density.

(The validity of this relation for a range of n-alkanes means no more than that their Raman LA lowest mode (accordion mode) has a frequency inversely proportional to the number of carbon atoms in the chain and this is true only for the crystalline material).

Modifications of the "rod" model to account for end conditions in, for example, lamella crystals do not seem to be worthwhile except as curve-fitting exercises, for the "rod" can have no physical meaning.

If identification of lattice vibrations with those of a continuum is made as in Born's theory then frequencies even lower than the Raman LA mode would be expected from the extreme anisotropy of the polymer crystal. This may in fact be the case, but to assign continuum properties to the "rod" used in the accordion mode model is not likely to be fruitful.

Shimanouchi et al.[105] also derive a set of force constants from the Raman LA mode data by analysing the normal modes of vibration of the alkane chains. These constants are then used to predict the chain modulus using a simple model for the planar zigzag chain.

The differences between McCullough's approach and that of Shimanouchi is therefore the use of IR force constants in the first and Raman force constants in the second. In both analyses an effective E approaching 380 GPa is found after due correction for density/cell dimensions.

A different approach, but one which still requires the values of force constants to be supplied was made by Odajima and Maeda[107] using the "method of long waves" of Born[108] and the force constants of Schachtschneider and Snyder[109]. This method (which was also used later by Wobser and Blasenbrey[110]), requires the knowledge of the lattice constants and setting angle of the chains in the lattice as well as the assumption of a perfect and infinite crystal. The basis of the lattice dynamics approach is as follows. The potential energy Φ is made up of the intramolecular energy (the energy along the chain) and intermolecular energy (between chains) each summed over all the atoms in the unit cell. Odajima & Maeda and Wobser & Blasenbrey both used the IR data of Schachtschneider & Snyder but the former workers summed Φ over nearest neighbours only whereas Wobser & Blasenbrey extended the sums to second nearest neighbours.

The second derivative $\Phi_{\alpha\beta}(ij)$ of the potential energy Φ, for a given pair of particles i, j, is a 3×3 matrix whose $\alpha\beta$ element represents minus the force in the α direction on particle i when particle j is displaced in the β-direction.
The equation of motion of the ith particle is therefore

$$m_i\ddot{u}_\alpha(i) + \sum_{\beta,\,j} \Phi_{\alpha\beta}(ij)u_\beta(j) = 0$$

The sets of equations are solved by the assumption of periodic waves and, by expansion in powers of the wave number, a relation is found for the limiting case of long waves so that the elements of the dynamical matrix $\Phi_{\alpha\beta}$ can be related to the elastic constants of the continuum. It is also possible to derive the Raman frequencies from the lattice dynamics analysis but this does not seem to have been done for polymer crystals, though they have been derived for example, for NaCl and for diamond.

Two assumptions are made in Born's theory: 1) every nucleus is in equilibrium 2) the configuration corresponds to vanishing stresses. In addition, of course, it is assumed that there is a lattice, and that valid potential energy functions are available. It is not therefore possible, as McCullough pointed out, to use the methods of lattice dynamics in the case of disordered crystals – ones containing folds, kinks or jogs. For this reason McCullough et al. calculated chain stiffnesses for three basic types of conformational building blocks only one of which, the all-trans configuration has usually been considered in calculations of moduli at the present time.

6.1 The Microstructure of Polymers

In addition to the studies of chain configurations by McCullough et al. there has been considerable effort in studying detailed paracrystalline models by the Hosemann school[111] and in some recent work by Falk[112], an attempt to include microstructure into a theory of elasticity for polyethylene. In this work strain and structure curvature tensors are defined and related to dislocation and disclination content, thereby enabling such factors as chain bending and twisting to be considered. A conclusion of this theory is that surface forces *alone* cannot, if acting perpendicular to the chains, cause homogeneous deformation either in tension or in shear, for they will cause twisting of the surface chains.

The present authors believe that there are also problems to be considered when forces are transferred to a polymer crystal in directions parallel to the chains for the following reasons. In reality forces are transferred from atom to atom, primarily along the chains, not between chains.

The rationale for this view is the very large dependence (at all events in polymers such as polyethylene) of the overall moduli (calculated for example by the lattice dynamics approach) on the intramolecular force field. The calculated Young's moduli along and perpendicular to the chains are in a ratio of about 30 : 1 for polyethylene. Reverting for the sake of a physical model to the continuum model in which a chain behaves as a rod of high anisotropy it has been shown theoretically and by experiment (see Chaps. 2 and 3) that end effects in such highly anisotropic rods persist for very large distances away from the end. Since polymer crystals in a melt form part of a complex network of tie chains, folds and crystalline regions it is probable that load transfer through the crystalline regions is such that nowhere is there uniformity of stress (in the meaning of force divided by unit area of crystal). To assume a continuum model in calculations of overall elastic modulus in real crystals is therefore highly speculative at least until studies of load transfer in polymer crystals have been carried out for example by numerical analysis.

Measurements of strain using X-ray diffraction can be made on polymers but to derive a modulus from these requires the assumption that the applied stress is known at the crystal level. This for the reasons given above, is not valid in the present state of polymer physics.

7 Conclusion

In this review we have outlined those parts of classical elasticity and continuum mechanics which are relevant to experimental test methods and have then described the methods commonly used in polymer science with particular emphasis on the problems encountered in experiment.

a) We have demonstrated the importance, particularly with anisotropic materials, of "end-effects", that is effects associated with the application of load to the sample, whether by clamps or by point contact.

The effects of misalignment, again particularly with anisotropic materials, have been discussed and we have described how these effects may be detected and avoided.

b) We have discussed the problem of extrapolating from an experimental measurement to a general statement about the material. The relevance of the scale of the sample to that of the representative volume element (RVE) is stressed and some calculations and estimates of the size of the RVE are given for various polymers.

c) A review has been given of calculations of the elastic properties of polymer chains available in the literature and we have considered the applicability of these in the context of b) above.

d) Models of structure have been listed and their usefulness as real structural elements considered. It has been stressed that in some cases models are useful only as an aid to curve-fitting (although this is not without its uses) but in other cases the models have real structural significance and enable part of the bridge between molecular and continuum concepts to be rigorously constructed.

8 References

1. Love, A. E. H.: A treatise on the mathematical theory of elasticity. 4th Edit. Dover Press, New York 1944
2. Green, A. E., Adkins, J. E.: Large elastic deformations. Clarendon Press, Oxford 1970
3. Sokolnikoff, I. S.: Mathematical theory of elasticity. 2nd Edit. McGraw-Hill, New York 1956
4. Eringen, A. C. (ed.): Continuum Physics. Vol. 4, Academic Press, New York 1976
5. Saint-Venant, B.: Mémoires de l'académie des sciences des savants étrangers *14*, 233 (1855)
6. Pochhammer, L.: Z. Reine Angew. Math. *81*, 33 (1876)
7. Chree, C.: Trans. Camb. Phil. Soc. *14*, 250 (1889)
8. Filon, L. N. G.: Phil. Trans. Roy. Soc. *A 198*, 147 (1902)
9. Lur'e, A. I.: Three dimensional problems of the theory of elasticity. Intersci. New York 1964
10. Toupin, R. A.: Arch. Rat. Mech. Anal. *18*, 83 (1965)
11. Horgan, C. O.: J. Elast. *2*, 335 (1972)
12. Folkes, M. J., Arridge, R. G. C.: J. Phys. D: Appl. Phys. *8*, 1058 (1975)
13. Arridge, R. G. C. et al.: J. Mater. Sci. *11*, 788 (1976)
14. Fama, M. E. D.: Q. J. Mech. Appl. Math. *25*, 479 (1972)
15. Vendhan, C. P., Archer, R. R.: Int. J. Solids Structures *14*, 305 (1978)
16. Arridge, R. G. C., Folkes, M. J.: Polymer *17*, 495 (1976)
17. Lekhnitskii, S. G.: Theory of elasticity of an anisotropic body. Holden-Day, San Francisco, 1963
18. Hearmon, R. F. S.: An introduction to applied anisotropic elasticity. Oxford Univ. Press, London 1961
19. Raumann, G.: Proc. Phys. Soc. *79*, 1221 (1962)
20. Ladizesky, N. H., Ward, I. M.: J. Macromol. Sci. *B 5*, 745 (1971)

21. Biot, M. A.: J. Appl. Phys. *10*, 860 (1939)
22. Pipkin, A. C., Rogers, T. G.: J. Appl. Mech. *38*, 634 (1971)
23. Spencer, A. J. M.: Deformations of Fibre-reinforced materials. Clarendon Press, Oxford 1972
24. Thomson, W. T.: Vibration theory and applications. George Allen and Unwin, London 1965
25. Timoshenko, S.: Vibration problems in Engineering. 2nd Edit. Van Nostrand et al., New York 1937
26. Ritchie, I. G. et al.: J. Phys. D: Appl. Phys. *8*, 1733 (1975)
27. Hertz, H.: Z. f. Math. (Crelle) *92* (1881)
28. McCrum, N. G.: Priv. comm.
29. Hashin, Z.: Int. J. Engng. Sci. *7*, 11 (1969)
30. Laws, N., McLaughlin, R.: Proc. Roy. Soc. *A 359*, 251 (1978)
31. Lee, E. H.: Q. Appl. Math. *13*, 183 (1955)
32. Takayanagi, M.: Proc. 4th Int. Conf. Rheology, Part 1. Intersci. New York 1965
33. Smith, J. B. et al.: J. Polym. Sci., Polym. Phys. Ed. *13*, 2331 (1975)
34. Arridge, R. G. C., Barham, P. J.: Polymer *19*, 603 (1978)
35. Ward, I. M.: Mechanical properties of solid polymers. Wiley, New York 1971
36. Wedgewood, A. R., Seferis, J. C.: Polymer *22*, 966 (1981)
37. Gibson, A. G., Davies, G. R., Ward, I. M.: Polymer *19*, 683 (1978)
38. Odell, J. A., Keller, A.: Polymer Engng. Sci. *17*, 544 (1977)
39. Cooper, J. W., McCrum, N. G.: J. Mater. Sci. *7*, 1221 (1972)
40. Gilmour, I. W., Trainor, A., Haward, R. N.: J. Polym. Sci., Polym. Phys. Ed. *16*, 1277 (1978)
41. Godovskii, Yu, K., Slonimskii, G. L., Alekseev, V. F.: Vysokomol. Soedin. *11 A*, 1181 (1969)
42. Zener, C. M.: Elasticity and anelasticity of metals. Chicago Univ. Press, Chicago 1965
43. Nowacki, W.: Dynamic problems of thermoelasticity. Noordhoff Internat. Publ. Groningen 1975
44. Nowick, A. S., Berry, B. S.: Anelastic relaxation in crystalline solids. Academic Press, New York 1972
45. Barham, P. J., Arridge, R. G. C.: Polymer *20*, 509 (1979)
46. Sander, B.: Gefügekunde der Gesteine. J. Springer, Vienna 1930.
47. Hill, R.: J. Mech. Phys. Solids *11*, 357 (1963)
48. Takayanagi, M.: Mem. Fac. Engng. Kyushu Univ. *23*, 1 and 41 (1963)
49. Van Fo Fy, G. K., Savin, G. N.: Mekh. Polim. *1*, 151 (1965)
50. Pickett, G.: Trans. ASME *664*, 176 (1944)
51. Arridge, R. G. C., Lock, M. W. B.: J. Phys. D: Appl. Phys. *9*, 329 (1976)
52. Smith, G. E., Spencer, A. J. M.: J. Mech. Phys. Solids *18*, 81 (1970)
53. Broutman, L. J., Agarwal, B. D.: Polymer Engng. Sci. *14*, 581 (1974)
54. Whitney, J. M.: J. Compos. Mat. *1*, 121 (1967)
55. Rosen, B. W.: Proc. Roy. Soc. *A 319*, 79 (1970)
56. Hashin, Z., Rosen, B. W.: Trans. ASME, J. Appl. Mech. *31*, 223 (1964)
57. Hashin, Z., Shtrikman, S.: J. Mech. Phys. Solids *11*, 127 (1963)
58. Walpole, L.: J. Mech. Phys. Solids *17*, 235 (1969)
59. Kröner, E.: Z. Physik *151*, 504 (1958)
60. Roscoe, R.: J. Mech. Phys. Solids *20*, 91 (1972)
61. Walpole, L.: J. Mech. Phys. Solids *14*, 151 (1966)
62. Hill, R.: J. Mech. Phys. Solids *12*, 199 (1964)
63. Eshelby, J. D.: Proc. Roy. Soc. *A 241*, 376 (1957)
64. Wunderlich, B.: Macromolecular Physics, Vol. 1. Academic Press, New York 1973
65. Arridge, R. G. C.: Continuum Models of Discrete Systems *4*. Brulin, O., Hsieh, R. K. T. (eds.) North Holland, Amsterdam 1981
66. Haward, R. N., Daniels, H. E., Treloar, L. R. G.: J. Polym. Sci., Polym. Phys. Ed. *16*, 1169 (1978)
67. Gaskell, P. H., Gibson, J. M., Howie, A. in: The structure of non-crystalline materials. Gaskell, P. H. (ed.) Taylor and Francis, London 1977, p. 181
68. Hendus, H., Illers, K., Ropte, E.: z. Kolloid z.z. Polym. *216*, 110 (1967)
69. Matsuo, M.: Japan Plast. *2*, 6 (1968)
70. Dawkins, J. V. in: Block Copolymers. Allport, D. C., Janes, W. H. (eds.) Appl. Sci. Publ. London 1973

71. Allport, D. C., Mohajer, A. A. in: Block Copolymers, Allport, D. C., Janes, W. H. (eds.) Appl. Sci. Publ. London 1973
72. Arridge, R. G. C., Folkes, M. J.: J. Phys. D: Appl. Phys. *5*, 344 (1970)
73. Gray, W., McCrum, N. G.: J. Polym. Sci. A-2, *7*, 1329 (1969)
74. Maeda, M. et al.: J. Polym. Sci. A-2, *8*, 1303 (1970)
75. Buckley, C. P.: J. Mater. Sci. *9*, 100 (1974)
76. Andrews, E. H.: Pure and Appl. Chem. *39*, 179 (1974)
77. Patel, J., Phillips, P. J.: Polym. Lett. *11*, 771 (1973)
78. Halpin, J. C., Kardos, J. C.: J. Appl. Phys. *43*, 2235 (1972)
79. Owen, A. J., Ward, I. M.: J. Mater. Sci. *6*, 485 (1971)
80. Raumann, G., Saunders, D. W.: Proc. Phys. Soc. *77*, 1028 (1961)
81. Ward, I. M.: Proc. Phys. Soc. *80*, 1176 (1962)
82. McCullough, R. L. et al.: Polym. Engng. and Sci. *16*, 371 (1976)
83. Seferis, J. C., McCullough, R. L., Samuels, R. J.: Polym. Engng. and Sci. *16*, 334 (1976)
84. Halpin, J. C., Kardos, J. L.: Polym. Engng. and Sci. *16*, 344 (1976)
85. Barham, P. J., Arridge, R. G. C.: J. Polym. Sci., Polym. Phys. Ed. *15*, 1177 (1977)
86. Arridge, R. G. C., Barham, P. J.: J. Polym. Sci., Polym. Phys. Ed. *16*, 1297 (1978)
87. Peterlin, A.: Copolymers, Polyblends and Composites: Adv. Chem. Ser. *142*, Am. Chem. Soc. 1975
88. Cox, H. L.: Brit. J. Appl. Phys. *3*, 72 (1952)
89. Peterlin, A. in: Ultra-high modulus polymers. Ciferri, A., Ward, I. M. (eds.) Appl. Sci. Publ. London 1979, p. 279
90. Ciferri, A., Ward, I. M. (eds.): Ultra-high modulus polymers. Appl. Sci. Publ. London 1979
91. Chen, F. C., Young, K.: J. Math. Phys. *18*, 1412 (1977)
92. Capaccio, G., Ward, I. M.: Polymer *15*, 233 (1974)
93. Gibson, A. G. et al.: J. Mater. Sci. *9*, 1193 (1974)
94. Odell, J. A., Grubb, D. T., Keller, A.: Polymer *19*, 617 (1978)
95. Zwijnenburg, A., Pennings, A. J.: Colloid and Polym. Sci. *259*, 868 (1978)
96. Barham, P. J., Keller, A. Polym. Lett. *17*, 591 (1979)
97. Smith, P. et al.: Polym. Bull. *1*, 733 (1979)
98. Smith, P., Lemstra, P. J.: J. Mater. Sci. *15*, 505 (1980)
99. Treloar, L. R. G.: Polymer *1*, 95 (1960)
100. Rasmussen, R. S.: J. Chem. Phys. *16*, 712 (1948)
101. McCullough, R. L., Eisenstein, A. J., Weikart, D. F.: J. Polym. Sci., Polym. Phys. Ed. *15*, 1837 (1977)
102. Lifson, S., Warshel, A.: J. Chem. Phys. *49*, 5116 (1968)
103. Williams, D. E.: J. Chem. Phys. *47*, 4680 (1967)
104. Mizushima, S., Shimanouchi, T.: J. Amer. Chem. Soc. *71*, 1320 (1949)
105. Shimanouchi, T., Asahima, M., Enomoto, S.: J. Polym. Sci. *59*, 93 (1962)
106. Shauffele, K. F., Shimanouchi, T.: J. Chem. Phys. *47*, 3605 (1967)
107. Odajima, A., Maeda, T.: J. Polym. Sci. *C15*, 55 (1966)
108. Born, M., Huang, K.: The dynamical theory of crystal lattices. Clarendon Press Oxford 1954
109. Schachtschneider, J. H., Snyder, R. G.: Spectrochim. Acta *19*, 117 (1963)
110. Wobser, G., Blasenbrey, S.: Kolloid Z. Polym. *241*, 985 (1970)
111. Hosemann, R.: CRC Crit. Rev. Macromol. Sci. 351 (1972)
112. Falk, F.: Colloid and Polymer Sci. *258*, 142 and 365 (1980)

Received November 3, 1981
M. Gordon (editor)

Secondary Relaxations in Glassy Polymers: Hydrophilic Polymethacrylates and Polyacrylates

Jan Kolařík

Institute of Macromolecular Chemistry, Czechoslovak Academy of Sciences, 162 06 Prague 6, Czechoslovakia

Although secondary relaxations (dispersions, transitions) in glassy polymers have extensively been studied within the last two decades, current understanding of their molecular mechanisms still remains incomplete and qualitative. At present, it is usually possible to identify the structural units, i.e. short sequences of main chains or parts of side chains, whose motions give rise to relaxation phenomena; however, it is difficult to draw conclusions on the type of molecular motions and the mechanism of energy storage, transfer and dissipation. Categorization and characterization of molecular motions underlying secondary relaxations are briefly reviewed to surface some topical problems as yet unsolved. Furthermore, elucidation und classification of the effects of low-molecular weight compounds on molecular mobility have been attempted for the first time. An overview of the author's dynamic mechanical measurements of hydrophilic polymethacrylates and polyacrylates is presented to illustrate the effects of polarity, geometrical constraints, comonomers, and incorporated diluents on molecular mobility. Concepts of dynamics and intensity of dynamic mechanical relaxations and other methods employed for the study of molecular motions are shortly surveyed.

1 Introduction 120

2 Secondary Transitions in Glassy Polymers and Methods of Their Determination 120

3 Dynamic Mechanical Relaxations in Glassy Polymers 124
- 3.1 Dynamics of Relaxations 124
- 3.2 Intensity of Relaxations 128
- 3.3 Characterization of Relaxations 130

4 Types of Molecular Motions in Glassy Polymers 131
- 4.1 Local Main-Chain Motions 131
- 4.2 Side-Chain Rotations About the Bonds Linking Side Chains to the Main Chain 133
- 4.3 Internal Motions within the Side Chain 133
- 4.4 Diluent-Induced Secondary Relaxation 134

5 Dynamic Relaxation Behavior of Hydrophilic Polymethacrylates and Polyacrylates in the Glassy State 136
- 5.1 Low-Temperature Relaxation 137
- 5.2 The β Secondary Relaxation 144
- 5.3 Effect of Diluents and the Diluent-Induced Secondary Relaxation 146

6 Conclusion 154

7 References 158

Advances in Polymer Science 46

1 Introduction

The mobility of main and side chains of macromolecules is one of the factors which determine the mechanical and other physical properties of solid polymers. The temperature dependence of molecular mobility is characterized by various transitions in which a certain mode of chain motion sets in (or freezes in with decreasing temperature). The most important transition in amorphous polymers or amorphous regions of crystalline polymers is the glass transition at the temperature of which, T_g, the micro-Brownian motion of segments of the main chains becomes active. The length of these segments is inversely proportional to the flexibility of the main chains; in common polymers it is estimated to be several tens of C–C bonds. The glass transition has received much attention in polymer physics because it is accompanied by significant changes in the mechanical (modulus of elasticity decreases by three or four orders of magnitude) and other physical properties of the polymer sample; all of them are important also with respect to its applications. The transitions which appear in glassy polymers are called secondary transitions. Since they are associated with the motions of short segments of main chains or with the motions of parts or the whole of the side chains, they are accompanied by much smaller changes in the physical quantities than the glass transition. Until now, secondary transitions have received relatively little attention, probably also because of their low practical importance so that the understanding of the molecular mechanisms involved is still incomplete, and the description of the observed phenomena is semiquantitative at best.

This review gives a brief characterization and classification of molecular motions giving rise to secondary transitions, an outline of the existing knowledge of the dynamics and intensity of relaxation processes observed in dynamic mechanical measurements, and an overview of our own dynamic mechanical studies of hydrophilic polymethacrylates and polyacrylates. The methods employed in the detection and investigation of secondary transitions are only briefly reviewed. From the vast and ceaselessly accumulating experimental results on the molecular mobility in glassy polymers, only some important representative examples have been selected. (A survey of experimental data and of the phenomenological theory, as well as details regarding molecular theories of the mechanical relaxation behavior can be found in a number of monographs[1–8)] and reviews[9–15)].) Special attention has been devoted to the elucidation and categorization of effects of low-molecular weight compounds on molecular mobility. Due to the generally known fact that the understanding of relaxation phenomena is still rather incomplete and mostly only qualitative, many of the conclusions presented in this review are merely tentative.

2 Secondary Transitions in Glassy Polymers and Methods of Their Determination

Secondary transitions in glassy polymers are closely associated with limited molecular mobility, i.e. with the rotational and vibrational motions of relatively short chain sections. The motional units may be identified with sequences of the main chains consisting of four to six groups, or with side chains and their parts. Generally, it is believed that such a motional unit may assume several stable conformations separated from each other by potential barriers; the frequency of jumps over a potential barrier is inversely proportional to its height and proportional to the absolute temperature. This is obviously a rate process similar to the rotational isomerization of low-molecular weight compounds. If a molecular motion is detected by using a dynamic method with variable frequency of the force field, energy is absorbed in the polymer at frequencies close to that of the motion. In dynamic mechanical experiments, the formation of the absorption maximum (peak of the loss modulus) is accompanied by a concomitant change in the modulus. It is obvious that the temperature location of an observed transition, which involves a conformational isomerization, must be supplemented by data on the frequency of measurement because at a higher frequency the same motion is detected at a higher temperature. Since absorption of energy is due to relaxation mechanisms, the term "secondary transition" is very often replaced by such expression as relaxation (relaxation region) or dispersion (dispersion region). The "onset" or "freezing-in" temperature of a certain local molecular motion is also related to the energy absorption phenomenon (this is the temperature location of a relaxation observed with increasing or decreasing temperature of measurement) and does by no means imply that an abrupt change occurs in the mobility of the motional unit under investigation (cf. Sect. 3.1.1). It has become customary to denote the relaxation regions as α, β, γ, δ in conformity with decreasing temperature of relaxation ($T_\alpha > T_\beta > T_\gamma > T_\delta$); α is reserved for the relaxation corresponding to the glass transition while δ is used for cryogenic relaxations at temperatures lower than the boiling point of nitrogen.

Investigation of the temperature dependence of thermodynamic variables has shown that in the range of secondary transitions the expansion coefficient α usually exhibits an S-shaped increase with rising temperature[16–18]; in some particular cases[19], a rise of isothermal compressiblility was also detected ($\Delta\beta$ was about 1×10^{-10} Pa^{-1} for poly(methyl methacrylate) and plasticized poly(vinyl chloride)). In several papers[19–22], an increase in the temperature of a secondary relaxation with rising hydrostatic pressure was observed which is in qualitative agreement with the equation $dT/dP = \Delta\beta/\Delta\alpha$ for second-order thermodynamic transitions. In no case, however, can the secondary transitions discussed here be regarded as second-order thermodynamic transitions because changes in $\Delta\alpha$ and $\Delta\beta$ occur within a wide temperature range and the shape of the curves does not fit that predicted by theory[2]. The corresponding change in the specific heat, ΔC_p, has not yet been observed, obviously because the equilibrium number of conformers varies with temperature very slowly without any abrupt change[18, 23] (Sect. 3.1). In this connection, it should be emphasized that in practical applications the polymeric glasses are never present in a thermodynamic equilibrium state[9, 10]. This equilibrium is approached within extremely long times and the resulting changes in structure and physical properties are termed as physical ageing[4, 24].

Of the thermodynamic quantities just mentioned, only the determination of the expansion coefficient or other quantities reflecting its change have assumed practical importance for the identification of secondary transitions in glassy polymers. The most efficient methods for the investigation of the dynamics and intensity of molecular motions have so far been those based on the interference between molecular motion and the oscillating magnetic, electric or mechanical force field. In recent years, methods which employ various probes or labels in the study of molecular mobility have increasingly been used.

Secondary transitions of glassy polymers can be studied by means of the following methods:

(a) Dilatometry[16–20], a method which, in many cases, is suited only for the identification of the secondary transition because the change in the linear expansion coefficient, $\Delta\alpha$, is small, usually varying between 1 and 3×10^{-5} K^{-1}. Since the increase by $\Delta\alpha$ proceeds in the range from 20 and 40 K, the transition temperature can be determined approximately as the position of the inflexion point.

(b) Infrared spectra[25–27] which vary with increasing expansion coefficient (i.e. in average distances between macromolecules) due to a decrease in the magnitude of absorption, while shifts of vibrational frequencies are usually negligible because of a small contribution from the intermolecular force field to the potentional energy for a particular vibration. For this reason, secondary transitions are detected as an abrupt change in the slope of the absorbance vs. temperature plot. Since the structural origin of the bands is known, the groups or structures contributing to each transition can be identified.

Thermally stimulated methods are based on the fact that a sample frozen in a deformed or polarized state gradually returns to equilibrium as various types of molecular motions set in with increasing temperature.

(c) Thermally stimulated (dielectric) depolarization[4, 28–30] which has been used in the study of secondary transitions in glassy polymers since the early seventies. A transition temperature at which the molecular motion of dipole-bearing groups is liberated is identified as the temperature of the depolarization current peak. By selecting a suitable temperature regime during the experiment[29], i.e. the polarization and depolarization temperatures, data may be obtained on the temperature dependence of relaxation times and their distribution.

(d) Thermally stimulated creep recovery,[31, 32] which is a mechanical analog of the preceding method; it is a recent technique employed in the study of relaxation processes. A fractional loading program along with an appropriate temperature regime has been developed so as to improve the resolution of the method and to determine the components of complex compliance within a broad frequency and temperature range. The data obtained allow the dynamics of molecular motions to be characterized by means of the spectrum of relaxation times or activation energies.

(e) Thermoluminescence[30, 33, 34], occurring when electrons released from traps recombine with the luminescence centers, can be used in the detection of transitions, assuming that the onset of a certain molecular motion leads to an escape of electrons from a certain type of traps and to a maximum of glow discharge. The interpretation of the results is however complicated because one has to identify traps in which electrons are captured when the sample is irradiated with gamma or UV rays (at low temperatures), luminescence centers, effect of chemical additives (catalysts, antioxidants, etc.), and many other factors. This is why the method has not been applied on a wider scale, and is mentioned in this survey rather for historical reasons.

Dynamic methods rank among those most frequently used for molecular mobility studies in polymers. Their principles and application have been described in detail[1, 4–8]. The dynamic methods mentioned below differ in their selectivity, measuring frequency and frequency range covered. The results obtained by these methods can be combined if molecular mobility is to be studied in greater detail.

(f) Nuclear magnetic resonance[4, 35, 36] which is mostly represented by proton magnetic resonance because in ordinary polymers only the hydrogen atom has a non-zero nuclear spin. The ^{13}C high-resolution nuclear magnetic resonance has recently also been utilized[37]. Both techniques, i.e. measurement of the wide-line narrowing and of the spin-lattice relaxation times, are suitable for the investigation of the dependence of molecular mobility on temperature. A disadvantage of the method may be seen in the high frequency of measurement (in the order of MHz) which leads to an extensive overlap of detected transitions with different activation energies. This method is suited, among other things, also for the determination of the onset of rotation of variously bound methyl groups as is the case with poly(methyl methacrylate)[1].

(g) Of the methods which employ probes (free "guest" molecules in the polymer matrix) or labels (molecules incorporated into or linked to the polymer chain by covalent bonding), electron spin resonance is the most important; during the last decade, it has proved to be a versatile method for investigating molecular transitions in various polymeric materials. Stable nitroxide radicals are usually employed as paramagnetic probes[39, 40] or labels[41]; their rotational and translational mobility, estimated from the ESR spectra, reflects the molecular mobility of the polymeric matrix. When analyzing the temperature dependence of the spectra, one should bear in mind the interaction between radicals and the polymeric matrix and decide to which kind of molecular motion the given probe radical (of various size and shape) could respond. In the present stage of development, the method can be regarded as appropriate for the detection of transitions in various polymer systems. The isomerization of azoaromatic groups as photochromic labels[42] or probes[43] as well as fluorescence[44, 45] and phosphorescence[46] of various labels or probes can also be included into this group.

(h) Dielectric measurements[1, 4, 7, 47] which can be performed within a very wide frequency and temperature range. This method provides relatively complete information on molecular motions based on relaxation phenomena. In order to raise the versatility of the method, permanent dipoles at small concentrations, sufficient for the indication of molecular motions, are introduced into nonpolar polymers.

(i) Dynamic mechanical measurements[1, 3–6, 8] which are the most universal of all the methods described before; this is because they can determine a change in the mobility of almost all types of motional units. The method is also applicable to polymers of low strength or insufficient dimensional stability if they are coated on a carrier (braid) of suitable mechanical properties (torsional braid analysis[48], sample – supportive techniques[49]). A shortcoming of many devices results from a comparatively narrow frequency range which necessitates a combination of the results obtained by several dynamic mechanical methods, if a frequency dependence of the dynamic mechanical response is to be constructed. As regards the amount of results collected so far, dynamic mechanical measurements predominate over the other methods. However, a deeper analysis and a quantitative interpretation of results are impeded by poorly established relations to molecular motions. In Chap. 3, we try to summarize briefly the present state of knowledge in this field.

3 Dynamic Mechanical Relaxations in Glassy Polymers

3.1 Dynamics of Relaxations

A macroscopic change in the shape of a solid, both reversible (elastic) and irreversible (plastic), is the sum of individual displacements on the atomic level. Atoms in the solid phase occupy equilibrium positions and vibrate about the minimum of free energy or enthalpy[50]. When the polymer sample is exposed to stress, the atoms are elastically shifted from their equilibrium positions; the strain energy is transformed into the potential energy of the system which increases and is stored reversibly. After the external stress has been removed, the atoms return to their equilibrium positions and the stored energy is recovered (mechanical and thermodynamical reversibility). In a general case, the time-independent elastic deformation may be accompanied by a time-dependent viscoelastic (mechanically reversible) or plastic (irreversible) deformation. The plastic flow or viscous component of the viscoelastic deformation are usually regarded as thermally activated processes in which the stable positions of atoms or of other structural units are separated from each other by potential barriers. In order to overcome the barrier, the motional unit must acquire sufficient energy. The average energy of the whole system is proportional to the absolute temperature, T, but at any definite temperature some motional units possess less energy then corresponds to the average while a limited number of units possess much more energy. Only these units with high energy are able to pass over the potential barrier to other stable positions (sites).

From the theory of absolute reaction rates it follows[4, 6, 50–56] for a simple barrier model (Fig. 1 a) that the rate constant of the jumps of a particle (motional unit) from site 1 to site 2 is given by

$$k_{12} = A \exp(-\Delta F_1/RT) , \qquad (1)$$

while for the opposite direction the relation

$$k_{21} = A \exp(-\Delta F/RT) \qquad (2)$$

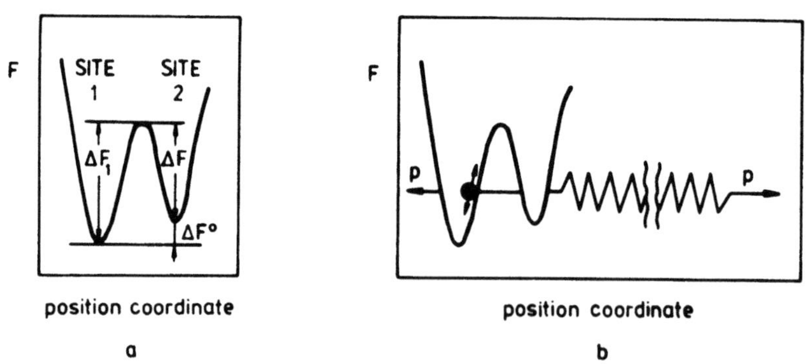

Fig. 1 a. Two-site model; **b** Schematic view of the effect of external force p acting on the motional unit in the two-site model

holds where $A = kT/h$ (k and h respectively are the Boltzmann and Planck constants), $\Delta F_1 = \Delta F + \Delta F^\circ$ is the activation Gibbs free energy, and R is the gas constant. If the number of particles is n, of which n_1 particles are in position 1 and n_2 in position 2, then $dn/dT = k_{12}n_1 - k_{21}n_2 = 0$ in the dynamic equilibrium so that

$$n_2/n_1 = m_2/m_1 = K = \exp(-\Delta F^\circ/RT) , \tag{3}$$

where

$$m_1 = n_1/n \text{ and } m_2 = n_2/n .$$

In the stress-free state, the occupation of both sites is determined only by the absolute temperature. The deformation of glassy polymers does not involve (for kinetic reasons) any flow or micro-Brownian motion of chain segments but only deformation of bond angles and bonds, an increase in the intermolecular distances and motion connected with rotations about the bonds[57–59]. Deformation of bond angles and bonds, though elastic, means high stress and high modulus which is observed, e.g. with uniaxially oriented polymers in the chain direction. Intermolecular forces (nonbinding interactions) yield low theoretical values of the modulus (decreasing markedly with increasing distance between chains during deformation) which suggests that they control the modulus of glassy polymers only to a minor degree[57].

A strained glassy polymer may reduce the resulting stress by populating other local conformations through rotational motions of side chains and short sections of main chains. This situation is schematically represented (Fig. 1 b) by a simple site model (cf. Ref. 60) where the force p (per motional unit) acts upon the unit in site 1 via a linear spring (representing purely elastic modes of deformation) and raises the probability of a thermally activated jump of the unit into site 2 which accounts for a decrease in the stress of the spring and a decrease in the rigidity of the model. The effect of the acting external stress is also described[6, 9, 11] as a rise in the potential energy in site 1 and its decrease in site 2; the decline in ΔF° is then reflected in an increased population of conformers with higher energy. Rearrangement of the system connected with the energy transfer from internal (vibrational) to external (rotational) degrees of freedom is, however, time-dependent. In other words, all degrees of freedom with the exception of the conformational ones remain in equilibrium with the instantaneous temperature, pressure, and stress. The assembly of conformers eventually possesses higher energy than the initial one, because external work has been transformed into conformational energy. A model relaxation experiment, in which stress resulting from steplike deformation relaxes with time owing to increasing population of higher-energy conformers, would be impractical for glassy polymers. A cyclic perturbation of the equilibrium (not necessarily stress-free) state by an oscillating external force field is much more suitable, because in this way one can achieve a stationary state of molecular rearrangements during the measurement. Thus, the results of dynamic mechanical measurements reflect the dynamic relaxation behavior of polymers.

The rate of relaxation at constant temperature[51, 52] is generally proportional to the magnitude of displacement from the equilibrium and is characterized by a relaxation time τ:

$$-\,dm_2/dt = (m_2 - m_2^\circ)/\tau\ , \tag{4}$$

where m_2° is a quantity corresponding to the state of equilibrium. (In contrast to resonance phenomena, in relaxation phenomena the inertial force connected with the mass of the motional unit is not operative). Using the relations $K = k_{12}/k_{21}$ and $m_2 = 1 - m_1$, the relation $-\,dm_2/dt = k_{21}m_2 - k_{12}m_1$ can be rearranged to

$$-\,dm_2/dt = k_{21}(1 + K)[m_2 - K/(1 + K)]\ . \tag{5}$$

By comparing Eqs. (4) and (5), we obtain the following dependence of the relaxation time on temperature:

$$\tau = [k_{21}(1 + K)]^{-1} = (1/A)\exp(\Delta F/RT)[1 + \exp(-\,\Delta F^\circ/RT)]^{-1} \tag{6}$$

Thus, the relaxation time of a certain molecular response to perturbation is expressed in the parameters of the stress-free equilibrium state. If RT is much smaller than ΔF°, the approximation[6]

$$\tau = \tau_0 \exp(\Delta F/RT) \tag{7}$$

can be used, where $\tau_0 = 1/A$.

Secondary transitions associated with rotational motions (formation of conformers) in glassy polymers are usually characterized by the activation enthalpy ΔH or by the activation energy ΔU and not by the activation Gibbs free energy $\Delta F = \Delta U + P\Delta V - T\Delta S$, because it is very difficult to determine the corresponding changes in volume, ΔV, and entropy, ΔS. Since these transitions are not of the order-disorder type[4, 10, 61], the entropy of the systems is regarded as constant, and $\Delta S = 0$. Recent investigations carried out in great detail[62] show, however, that ΔS of the secondary transitions in poly(vinyl chloride) and poly(methyl methacrylate) attains values of about 0.04 kJ/mol K so that for relaxations at room temperature the contribution $T\Delta S$ amounts to as much as 12 kJ/mol (while ΔU is about 72 kJ/mol). The contribution $P\Delta V$ for the secondary transition associated with the rotation of the side chains in poly(methyl methacrylate)[63] is some 2 kJ/mol at 100 MPa; it is obvious, therefore, that at normal pressure it can be neglected.

A great number of dynamic mechanical and dielectric measurements have supplied sufficient evidence that τ_0 may be regarded as a constant virtually independent of the temperature of measurement and of the polymer composition. If simplifying assumptions $\Delta S = 0$ and $\Delta V = 0$ are made, the Arrhenius equation is obtained

$$\log \tau = \log \tau_0 + 0.4343\,\Delta U/RT\ . \tag{8}$$

It is mostly used to describe the effect of temperature on the dynamics of relaxation processes in glassy polymers. Rearranging Eq. (8), we obtain the relation between the activation energy ΔU (in kJ) and the temperature of relaxation, $T_r(f)$, detected at frequency f:

$$\Delta U = (K_1 - 0.0191 \log f)\, T_r(f)\ . \tag{9}$$

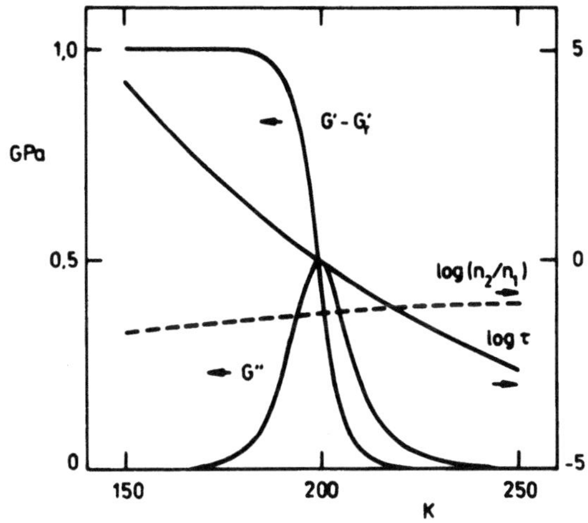

Fig. 2. Temperature dependence of the storage modulus G′, loss modulus G″, relaxation time τ and ratio (n_2/n_1) of the equilibrium numbers of conformers for a single relaxation time model

By evaluating an extensive set of experimental data, an average value of $K_1 = 0.251$ kJ/mol was acquired[12, 13] which corresponds to an average value of $\log \tau_0 = -13.92$ (values 13.5 ± 1 are usually given). Equation (9) is very useful as it makes possible to check the experimental data or to estimate ΔU of the relaxation for which no experimentally determined values are available.

The intensity of relaxation phenomena observed in dynamic mechanical measurements of glassy polymers depends on the relative magnitude of the relaxation time τ of a certain molecular motion and of the period T_f of the oscillating force field (Fig. 2). If $\tau \gg T_f$, rotational motions inside the chains are unable to follow oscillations of the force field; the unrelaxed modulus of the polymer, G_u, is high, and all the stored energy is returned without loss. If $\tau \ll T_f$, conformers with the higher energy are populated with negligible delay after the imposed stress and the stored energy is also recovered without loss; due to their formation, however, the relaxed modulus G_r is smaller than G_u. If τ and T_f are close to each other, the population of conformers lags behind the oscillating stress, resulting in a phase shift between stress and the produced strain. The delayed response of the polymer can formally be separated into two components: the energy stored and returned in phase and the energy returned out of phase (i.e. with a phase lag $\pi/2$) which is dissipated in the form of heat. Hence, the dynamic mechanical properties are generally described by the complex modulus $G^* = G' + iG''$, where the storage modulus G′ and the loss modulus G″ correspond to the stored and dissipated energies. Thus, the losses in mechanical energy may be ascribed to its storage and transfer between various degrees of freedom. An essential feature of this concept, which has been elaborated in detail for the determination of conformational transitions in liquids by means of ultrasonic measurements[51–56], is that conformers with higher energy must exist and be accessible during perturbation.

A coincidence between τ and T_f, accompanied by the formation of a maximum of energy losses (peak of the loss modulus G″ or of the loss factor $\tan \delta = G''/G'$), can be achieved by varying either the period T_f at constant temperature (isothermal experiment) or τ (by changing the temperature) at the constant period T_f (isochronous experiment). The latter procedure is experimentally easier to implement and is therefore more frequently used.

Figure 2 shows the temperature dependence of log τ according to Eq. (8), the ratio of the equilibrium populations of conformers with a higher (n_2) or lower (n_1) energy according to Eq. (3), and the dependence of the storage (G′) and loss (G″) moduli ensuing from the relation for the single relaxation time model[1]:

$$G'' = \Delta G\omega\tau/(1 + \omega^2\tau^2) \, , \tag{10a}$$
$$G' = G_r + \Delta G\omega^2\tau^2/(1 + \omega^2\tau^2) \tag{10b}$$

where $\Delta G = G_u - G_r$. The initial values of the model calculation are: $\omega = 1\ s^{-1}$; ΔG = 1 GPa; ΔF° = 5 kJ/mol and ΔU = 50 kJ/mol which, according to Eq. (9), correspond to T_r = 199 K (1 Hz). Figure 2 shows that in the temperature range of dispersion there is no abrupt change in the relaxation time or in the population of the conformers.

The secondary transitions of real polymers exhibit a broad distribution of relaxation times which originates from the distribution of activation energies[1, 29, 64–67] (while in the glass transition range[1, 68], τ_0 is the distributed quantity). Hence, a change in temperature means a change in the distribution of relaxation times which causes considerable difficulties in the analysis of the isochronous temperature dependences of moduli. By using simplifying assumptions, however, it is possible to obtain semiquantitive information from these measurements on the average activation energy values, distribution of relaxation times and extent of relaxation processes[1, 64, 65]. A broad distribution of relaxation times or activation energies results in a sizeable broadening of maxima of energy losses and mutual overlapping of adjacent maxima. Since the activation energy of a relaxation process is proportional to its temperature according to Eq. (9), a frequency rise in the isochronous measurement means a poorer resolution of relaxations because a shift of loss maxima towards higher temperatures is the larger, the lower the activation energy (and the temperature of the relaxation process).

A deeper physical insight into the character of secondary relaxation processes is offered by a series of isothermal measurements within a broad frequency range. For this purpose, results of several dynamic mechanical methods are combined[12, 13, 69–73] because the individual methods usually do not cover the necessary frequency range. The use of the time-temperature superposition, which is very practical and widely employed in the main transition range[1, 68], is rendered more difficult by the temperature dependence of the spectrum of relaxation times of secondary transitions. The frequency dependences of moduli determined at various temperatures can form a generalized curve by using horizontal shifts (along the frequency axis) derived from the Eyring rate theory and empirical vertical shifts (along the axis of the modulus)[73]. The unsolved physical problems related to such superposition are the reason why it has been rarely used (cf.[62, 74, 75]).

3.2 Intensity of Relaxations

The site model appears to provide an adequate description of basic features of the dynamics of molecular motions in secondary transitions. On the other hand, however, attempts to calculate and interpret the extent of energy losses, i.e. the relaxation strength, in terms of the barrier model have so far been less successful. Theories employing a model with the same depth of potential energy minima[1, 76, 77] do not seem suitable because they do not take into account the storage of mechanical energy (they may however be used e.g. for the description of plastic flow). Theories[6, 9, 78] based on a model with the energy difference $\Delta F°$ between various sites use for the relaxation strength the expression

$$\Delta G = G_u - G_r = (N\,b^2/2\,RT)[\cosh^2(\Delta F°/2\,RT)]^{-1}\ , \tag{11}$$

in which b is the "mechanical dipole" and N the number of "dipoles", i.e of motional units, in a volume unit. This theory was subject to extensive experimental testing on polymethacrylates with saturated rings in the side chains[11] and was found to fail in describing the temperature dependence of the relaxation strength. The prediction of the theory that the relaxation strength is largest for $\Delta F° = 0$ is of course at variance with the theory of thermal relaxation of liquids[51–56].

Measurements carried out within a broad frequency range showed[8, 11–13, 69, 70] that results of ultrasonic measurements on glassy polymers in the range of secondary transitions are a continuation of those obtained by dynamic mechanical measurements at sonic and infrasonic frequencies. This continuity of the experimental data was observed both for the storage and for the loss modulus. There is no doubt, therefore, that both types of methods indicate the same molecular motions. Furthermore, ultrasonic measurements reveal an analogy between the rotational isomerization in liquid low-molecular weight compounds and glassy polymers, particularly in polymethacrylates with saturated six-, seven- und eight-membered rings in the side chains[11, 55, 56, 70, 79]. In ultrasonic measurements, however, the perturbation of the molecular motion is due to cyclic compression and decompression which causes a cyclic increase or decrease in temperature (under adiabatic conditions, this value amounts[54] to $10^{-3} - 10^{-2}$ K) and, consequently, changes in the population of conformers (according to Eq. (3)). A similar behavior of the molecular motions underlying the secondary relaxations of polymethacrylates with saturated cyclic substituents and isomerization of these rings in the liquid state is documented by the fact that in spite of the different ways of perturbation and different frequencies of measurements, the relaxations are characterized[11, 13, 70, 80] by the same values of ΔU or $\Delta U°$.

The theory of thermal relaxation of liquids[51–56] in the ultrasonic field, which has been worked out much better than the theory of relaxation processes in glassy polymers, postulates that oscillations of the translational energy of the liquid (owing to temperature oscillations) are accompanied by energy transfer into the other degrees of freedom until equilibrium is reached between all the energy modes (translational, vibrational, rotational and structural). Energy storage in the rotational degrees of freedom is due to changes in the population of conformers; as long as the energy transfer lags behind temperature fluctuations, losses in energy and attenuation of the ultrasonic wave take place. The relaxation strength is proportional to the relaxing part of specific heat which is given[51] by

$$\Delta C_{pr} = R(\Delta H°/RT)^2 \exp(-\Delta H°/RT)/[1 + \exp(-\Delta H°/RT)]^2\ . \tag{12}$$

This equation implies that the relaxation strength of a molecular motion passes through a maximum at $\Delta H° = 2.4$ RT and that for $\Delta H° = 0$ no relaxation process can exist. An equation for the loss factor $\tan\delta$ derived from this theory was attempted[81] for the calculation of the relaxation strength of the secondary process of poly(methyl methacrylate) (assigned to the hindered rotation of methoxycarbonyl groups). The analogy was believed to be justified by the fact that the extent of mechanical losses corresponds to $\Delta H^\circ = 14.2$ kJ/mol, which is in reasonable agreement with a value obtained from ultrasonic measurements for liquid ethyl acetate as a model low-molecular weight compound.

In a thermodynamic theory[82, 83)], the secondary mechanical relaxations are regarded as resulting from energy transfer between strain-sensitive (intermolecular) and strain-insensitive (intramolecular) modes of molecular motions in polymers. According to this theory, energy is stored in intermolecular modes in the course of straining and then transferred to intramolecular modes. The low-temperature relaxation process in poly(methyl methacrylate) tentatively attributed to the onset of mobility of methyl groups has been interpreted in terms of an energy exchange between the rotation of methyl groups and vibrations of the main chains. The theory has not been further generalized and confronted with other experimental data.

An alternative of the site model is the local mode relaxation model[84–86)] which assumes damped oscillations of chain segments about their equilibrium positions without jumps over potential barriers. It is hypothesized that torsional vibrations, which usually have a low force constant, are excited by the oscillating external force field to large amplitudes, but because of collisions with the adjacent chains their oscillations are damped. The local mode relaxation mechanism is identified with relaxation of the distribution function of the torsional displacements which approaches the equilibrium value after external excitation. This model is used only for polymers without side chains and is regarded as appropriate for the description of severely limited molecular motions in crystalline regions. Using molecular parameters, the drop in modulus ΔG in the relaxation region was calculated for several polymers and found to be in a reasonable agreement with experimental data. However, the model has not been sufficiently refined to make possible an adequate description of the dynamics of relaxation phenomena characterized by energy losses (cf.[1)]).

3.3 Characterization of Relaxations

By combining the results of several methods (dynamic mechanical, dielectric, NMR, etc.), it is usually possible to determine quite reliably the structural units whose motions give rise to secondary relaxations. If dynamic mechanical measurements alone are employed, the usual procedure is that the chemical constitution is systematically altered and correlated with the dynamic mechanical response spectra, i.e. with the temperature-dependence of the G'' and G' moduli. If the presence of a certain group in polymers is marked by the formation of a loss peak characterized by a certain temperature position, size and shape etc., then the conclusion may be drawn that the motional units responsible for the secondary relaxation are identical or related with that group. Naturally, the relations obtained in this way are empirical and qualitative.

In most cases, however, the question to be answered is what the motion of the identified motional unit is like and how the mechanism of storage and stransfer of mechanical energy proceeds. For some relaxation processes, it has not yet been possible to decide the primary problem, i.e. whether the molecular motion occurs inside the potential valley or over the potential barrier. While in many papers the dynamics of molecular motions is more or less amply characterized, the determination of the relaxation strength, its temperature dependence and related changes in the thermodynamic quantities have so far been quite exceptional. The question which of the existing theories, if any, adequately describes the character and extent of dynamic relaxation processes in glassy polymers has not yet been answered, mainly because set of the necessary experimental data is virtually lacking. The common routine measurements, mostly isochronous, of dynamic mechanical properties have as yet led to an incomplete description of molecular mobility for a number of polymers. In general, the dynamics of a particular molecular motion is usually characterized by (some of) the following quantities: the temperature of the peak of the loss maximum on the temperature axis and the corresponding relaxation time (frequency of measurement); activation energy, which controls the temperature dependence of the relaxation time, determined either by plotting the relaxation time against reciprocal temperature (according to Eq. (8)) or from the loss maximum area and the concomitant drop in the storage modulus[1,64,65]; distribution of relaxation times (activation energies) or parameter whose introduction into Eqs. (10a, 10b) with a single relaxation time permits to account for the distribution of relaxation times, e.g. the parameter of the Cole-Cole, Fuoss-Kirkwood equations etc.[1,64,65]. The extent of molecular motion is usually expressed as a change in the modulus, $\Delta G = G_u - G_r$, or as the loss peak area if the loss modulus (or, less suitably, the loss factor) is plotted against the reciprocal absolute temperature. It should be added that even incomplete dynamic mechanical measurements have contributed to the progress made at least in the qualitative understanding of the character of molecular motions responsible for secondary relaxations.

4 Types of Molecular Motions in Glassy Polymers

Molecular motions underlying secondary relaxation processes are a function of the constitution and structure of the polymer, but it is possible to find groups of polymers, usually of a similar composition, which exhibit an analogous transition (or transitions) characterized by similar values of temperature, activation energy, relaxation strength, etc. Molecular motions which give rise to secondary dispersions in glassy polymers above the liquid nitrogen temperature have been tentatively divided[12,13] into four types. This classification is also used in this paper, with only a brief description of the most typical relaxations; it is very likely that some other types may be met in practice. Strongly constrained molecular motions which occur at cryogenic temperatures[87–93] lie beyond the scope of this review.

4.1 Local Main-Chain Motions

This is the motion of short sections of main chains which seems to be virtually independent of the free volume (density) of polymers. Such a local residual mobility is the only possible cause of secondary relaxations of polymers without side chains[12]. It manifests itself as secondary (β) and/or low-temperature (γ) relaxations of polyethylene[1, 2, 61, 94–99], polyamides[1, 18, 100–103], polyurethanes[1, 104], poly(ethylene terephthalate)[1, 105–109], epoxides[110–112], poly(vinyl chloride)[113–115], polyoxymethylene[116], etc. No satisfactory description of the molecular mechanism of the local main-chain motion has so far been reported[1]. Japanese authors[84–86, 113] presume torsional oscillations of short segments inside the potential valley but in most hypotheses[14, 61, 97–99, 117–120] some variations of local conformations of the polymer backbone are supposed to proceed, being generally regarded as rotations about the main chain bonds, although small changes in the bond angles and bonds may also be assumed. An essential feature of the existing models is the colinearity of chemical bonds at the ends of the backbone sequences in question (Fig. 3). Several models of the local main-chain motions have been suggested:

(a) If the first and seventh bonds of the methylene sequence lie on the same axis, the intervening atoms – in the configurational sequence TGTGT (T = trans, G or G′ = gauche) – may rotate as crankshaft without affecting further atoms in the stems[121, 122]. In general, it may be objected that the volume swept out by the rotating section is too large to be readily accommodated in the condensed phase. It has also been pointed out that this model is not favored for statistical reasons[123]. Conformational analysis of polyethylene showed that this type of the crankshaft motion could hardly take place in the glassy matrix without extreme demands on energy[61].

(b) A simpler model of the crankshaft motion assumes rotation of two carbon atoms about the (coaxial) first and fifth bonds[124]. However, the energetically unfavorable cis-conformation of the central bond renders this model still less plausible than that described in (a).

(c) Another model of the low-temperature local motion of polyethylene[2, 125] assumes rotation of three carbon atoms about the colinear first and sixth bond. Although the model apparently requires a lower conformation energy and free volume, it has not been analyzed in greater detail.

(d) If the first und fifth bonds of the backbone are parallel to each other, but only approximately colinear, the internal bond sequence TGT may easily be transformed into TG′T. The energetics of such a conformational change is primarily determined

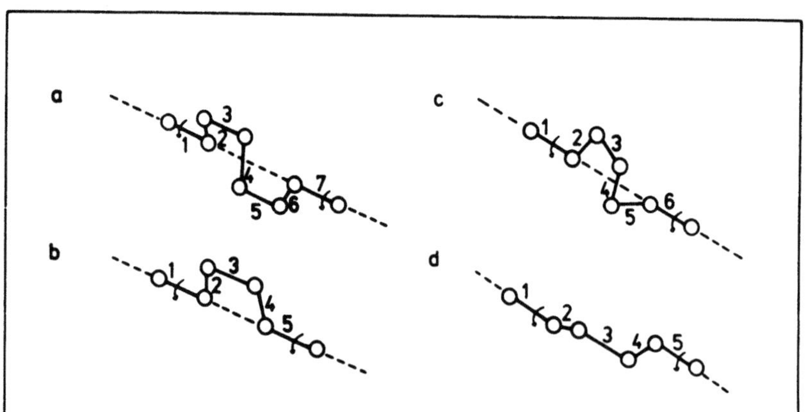

Fig. 3. Illustration of various types of crankshaft motions (cf. text)

by the rotational barriers of the bonds; no large distortions of bond angles and bond lengths are required. The conformational analysis[61, 97)] has brought evidence that this motion has to be given a serious consideration as a cause of relaxations in amorphous and crystalline domains of polyethylene, and also of some similar chemical structures.

The related kink model[119, 120)] is represented by the conformational sequence TGTG'T; the stems are slightly deformed in order to be situated on the same axis. The "flip-flop" kink reorientations TGTG'T ⇄ TG'TGT were believed to underly the low-temperature relaxation process in the crystalline regions of polyethylene. However, an energetics approach[98, 99)] has shown that the kink reorientations can only slightly contribute to the relaxation mentioned above, even though the associated energy barrier is consistent with experimental data.

In conclusion, one cannot but state, that the present-day knowledge of the mechanism of the low-temperature relaxation of polyethylene remains limited and qualitative, even though theoreticians have mainly studied this kind of molecular motion. The low-temperature relaxations of the other polymers without side chains are ascribed to analogous types of motion because the existing experimental data do not allow a better founded interpretation.

4.2 Side-Chain Rotations About the Bonds Linking Side Chains to the Main Chain

A typical representative of this category is the secondary relaxation of side chains COOR of polymethacrylates[1, 11, 12, 126, 127)] (250 to 300 K; 1 Hz), polyacrylates[1, 11, 12, 126, 127)] (about 200 K; 1 Hz), and polyithaconates[79, 128)]. It is generally accepted that the side chains as a whole perform rotary motions between the sites (Fig. 1 a). The temperature location of this relaxation is more affected by substituents of the adjacent mers than by the group R. The question still remains whether and to what extent the backbones participate in the relaxation motion. Some experimental evidence seems to indicate that the rotational motion of COOR requires a slight deformation of adjacent bond angles of the backbone[1, 129)]. Therefore, we refer to this type of motion as to a side group motion with some cooperation of the main chain[12, 13)]. Although phenomenological aspects of this relaxation are well developed, its molecular basis and mechanism remain questionable. This unsatisfactory state is caused, among other things, also by the fact that attempts to work out a theoretical model of such relaxation and to correlate this model with experimental data have so far been only sporadic[1, 130–132)].

4.3 Internal Motions Within the Side Chain

In contrast to the motions described above, no participation of the main chain is assumed for this type of molecular motion. A typical example can be seen in the hindered rotation inside the group R of the side chains COOR of polymethacrylates[1, 12, 13, 15, 69, 70, 80)], polyacrylates[1, 80, 127)] and polyithaconates[79, 133)], and others[134)]. It is generally assumed that this motion can be depicted by means of a simple site model. Data collected for polymers with cycloalkyl side groups[12, 13, 79, 80, 133, 134)] clearly show that the underlying conformational transitions (e.g. ring inversion between two different chair conformers) are quite similar to those detected in low-molecular weight compounds. The temperature of the onset of hindered rotation of alkyl, halogenalkyl and hydroxyalkyl side groups increases with the volume and polarity of substituents[1,15]. Although these molecular motions obviously rank among the simplest, relatively little attention has been devoted to their qualification and to the elucidation of the effects of composition on their parameters.

4.4 Diluent-Induced Secondary Relaxation

This type of molecular motions in glassy polymers has been investigated least of all, and attempts to interpret it, albeit only qualitative, remain rather speculative and often contradictory. It is essential to realize that the motions of diluent molecules may set in either independently of the existing molecular mobility of the polymers (in the diluent-free state) or at the same time with the motions of some parts of the main or side chains. In the latter case, too, the dynamic mechanical response spectra are often modified due to the occurrence of a new diluent-induced (β_d) loss maximum. Its formation and enlargement proportionally to the diluent concentration is accompanied in many cases by the reduction and eventually disappearance of the secondary (low-temperature) relaxation process the motional units of which interact with the diluent molecules. Diluent secondary relaxations are not caused by a certain unique type of molecular motion, as this was the case with the preceding types of relaxations, but most probably several different categories are to be distinguished. Our tentative classification leaves the relaxations, due to the motion of (or taking place within) diluent molecules, as an extra group. The other types of diluent-induced relaxations, which are usually assigned to the motion of complex motional units consisting of diluent molecule(s) and of a group present in the mers, are classified according to the type of motion exerted by this group in the absence of the diluent.

4.4.1 The Motion of, or Taking Place Within, Diluent Molecules Dissolved in a Polymer

A typical loss maximum of this type was observed for poly(methyl methacrylate) containing caprolactam or derivatives of cyclohexane[12,13]. It is noteworthy[70] that in the latter case the relaxation induced by the cyclohexyl group present in the incorporated plasticizer and the secondary relaxation of poly(cyclohexyl methacrylate) or poly(cyclohexyl acrylate) are characterized by an identical temperature position, 190 K (1 Hz), and activation energy, 47.9 kJ/mol ($\Delta U = 47.7$ kJ/mol is reported for the chair-chair transition of cyclohexanol). Hence, it can be seen that the cyclohexyl ring inversion, which represents a specific molecular motion, is remarkably insensitive to the surrounding molecules.

Of the diluents known to affect the dynamic relaxation behavior of polymers in the glassy state, water has so far received the greatest attention. Many polymers, which in the dry state are lacking any secondary relaxation process at temperatures from 77 to 273 K, e.g. poly(methyl methacrylate)[135], polymethacrylamide[136], cellulose and its derivatives[137], collagen[138], polysulfones[139], poly(2,6-dimethylphenylene oxide)[139], and others, exhibit – after the incorporation of water – a secondary loss maximum usually between 160 and 200 K (another loss maximum of collagen, located at 260 K is probably related with the melting of loosely bound or separated water). With increasing water content the loss peak is enlarged and usually shifted to lower temperatures. Since similar water peaks have been observed for polymers of different constitution, the conclusion has been forwarded[137)] that the mobility of water molecules linked by hydrogen bonds accounts for the relaxations. The relaxation mechanism is then interpreted as an exchange of water molecules between the sorption sites; the decrease in the relaxation temperature with water concentration is regarded as a consequence of looser bonding of further water molecules which gradually occupy increasingly weaker sites. This may also explain the observation that the temperature of the water peak is lower for less polar polymers. Small, if any, substrate conformational changes are assumed to take place.

The extent of the secondary relaxation process induced by water in collagen or polymethacrylamide is so large that the participation of the polymer in the arising relaxation seems to be unavoidable. It may only be speculated[140)], however, that the β_d relaxation is the result of specific interactions which produce unique structures exhibiting a characteristic relaxation process. On the other hand, a general explanation must be plausible also for diluent relaxations of systems containing nonpolar components (either polymer of plasticizer), for which specific interactions of structures could scarcely be assumed, as e.g. in the systems poly(methyl methacrylate)-benzene[141)] and polystyrene-dibutyl phthalate[142)].

4.4.2 The Motion of Diluent Molecules Associated with Local Main-Chain Motions

The interaction between the diluent molecules and polar groups of the backbone of polymers without side chains, e.g. polyamides[100–103)], polyurethanes[104)] and aliphatic polycarbonates[141)], leads to the reduction of the existing γ relaxation process (local main-chain motions) and to the formation of a new (β_d) loss maximum at a higher temperature (about 200 K). The β_d loss peak of polycaprolactam grows in size with the diluent content (up to a certain limit) and is displaced toward lower temperatures, even though it holds that $T_{\beta d} > T_\gamma$ at all concentrations; at the same time, the γ peak is reduced while its temperature position remains virtually constant. The observed alterations of relaxation patterns may be qualified as a transformation[100)] of the γ process into the β_d process. The γ process arising in the crystalline regions of polycaprolactam[100)] into which the diluent (water) does not penetrate remains unaffected. The transformation of relaxation processes ceases when a concentration is reached at which there is one water molecule per two amide groups of noncrystalline fractions. Since sorption measurements[143)] have revealed that a water molecule is bound by two hydrogen bonds to two amide groups of polycaprolactam, the result just mentioned can be regarded as favoring the view that the β_d process is closely associated with the interaction between diluent molecules and amide groups. Upon incorporation of formamide or acetamide, the β_d loss peak appears at a higher temperature, i.e. 255 and 230 K, which indicates that the diluent codetermines the location of the diluent peak. Both the polarity and volume of diluent molecules are probably operative in this case. An alternative interpretation (cf. Ref. 1) of the secondary process of polyamides as a motion of sequences containing unbound or weakly bound amide groups (also in the dry polymer) seems unlikely because infrared thermal analysis[144] has proved for a number of polyamides and their copolymers that almost all -NH- groups form hydrogen bonds even at room temperature.

Dynamic mechanical response spectra of elastin[145] (insoluble protein of vessels and ligaments), poly(ethylene terephthalate)[141] and polycarbonate based on Bisphenol A (4,4′-dihydroxydiphenylmethane)[141] show that incorporated water brings about enlargement of the existing secondary loss peak and its displacement toward lower temperatures. In conformity with the latter result, the activation energy of the relaxation process of elastin decreases. So far, no detailed data on this type of relaxation have been collected so that the coparticipation of water in the molecular motion cannot be specified more accurately.

4.4.3 The Motion of Diluent Molecules Associated with the Internal Motion Within Side Chains

The relaxation process due to the motion of the side groups is transformed[15, 146–148] by diluents into the β_d process in a way formally resembling that operative in the preceding case. The corresponding motional unit probably consists[148] of the group R and diluent molecule(s). Data obtained for polymethacrylates, which so far appear to be the most complete, are discussed in Sect. 5.3.

4.4.4 The Motion of Diluent Molecules Associated with Side Chain Rotation

This type of molecular motion seems to occur less frequently than the preceding ones. The existing results indicate that it is probably more characteristic of polyacrylates[127, 136] than of polymethacrylates[149]. Fragmentary evidence of this relaxation motion obtained up to now is presented in Sect. 5.3.

5 Dynamic Relaxation Behavior of Hydrophilic Polymethacrylates and Polyacrylates in the Glassy State

Polymethacrylates and polyacrylates have extensively been studied from the viewpoint of relaxations occurring in the glassy state. Though a vast amount of information has been collected to date, even a qualitative interpretation of the relaxation phenomena on a molecular level often remains questionable. This situation exists despite some favorable circumstances, i.e. polymethacrylates are amorphous polymers with comparatively simple molecular motions and it is possible to alter systematically their constitution and prepare various model polymers.

As mentioned earlier, we usually encounter two characteristic secondary relaxations in polymethacrylates and polyacrylates (below the glass transition temperature) which are assigned to side-chain motions[1, 12, 13, 15]: The β relaxation due to partial rotation of the COOR groups with some cooperation of the main chain and the γ (low-temperature) relaxation due to internal rotation within the side groups R. Incorporation of a diluent gives rise to a new (β_d) relaxation, occurring typically between 200 and 120 K, whose molecular mechanism is more complex and so far not well understood.

In this section, we have attempted to summarize some general features of the subglass relaxations and to single out the factors by which they are or are not affected. We partly refer to our dynamic mechanical measurements[15, 65, 66, 127, 136, 146, 148–160] performed by means of a freely vibrating (at about 1 Hz) torsional pendulum[161] with digital record of the amplitudes and period of the oscillations. Our efforts have been concentrated on the study of a series of selected methacrylate and acrylate polymers (and copolymers) listed in Table 1. A substantial part of our work has been devoted to an analysis of the effects of low-molecular weight compounds on the molecular mobility in the glassy state. We believe that a review of the results collected to date can contribute to a better understanding of the nature of the relaxations and surface some general problems as yet unsolved.

5.1 Low-Temperature Relaxation

The γ relaxation process[15, 66, 152, 153] of the hydroxyethyl groups of poly(2-hydroxyethyl methacrylate) (PHEMA) (Fig. 4) is located at T_γ = 140 K (all the temperatures given in Section 5 are related to 1 Hz) which is approximately by 45 or 23 K higher than that of poly(n-propyl methacrylate) (PPMA) or poly(2-chloroethyl methacrylate)[69]. The activation energy of the γ process decreases in the same order as the T_γ's, assuming the values 45, 29.3 and 21.8 kJ/mol, respectively. (The activation energy for PHEMA was determined[65] only approximately from the temperature dependence of the storage G' and loss G'' shear moduli. According to empirical relation (9), the activation energy is 35 kJ/mol.) Since the van der Waals radii decrease[162] in the order CH_3, Cl, OH, it is obvious that the increase in the T_γ's and ΔE_γ's is due to a rising interaction between the polar end groups of side chains.

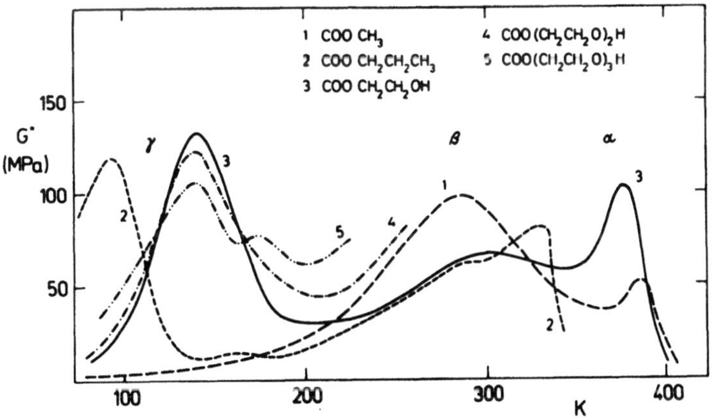

Fig. 4. Temperature dependence of the shear loss modulus of poly(methyl methacrylate) (*1*), poly(n-propyl methacrylate) (*2*), poly(2-hydroxyethyl methacrylate) (*3*), poly(5-hydroxy-3-oxapentyl methacrylate) (*4*), and poly(8-hydroxy-3,6-dioxaoctyl methacrylate) (*5*)

Table 1. Polymethacrylates and polyacrylates measured – $[CH_2\text{-}CR_1(COOR_2)\text{-}]_n$ and their relaxation temperatures

Polymer	Abbreviation	R_1	R_2	T_α^a (K)	T_β^a (K)	T_γ^a (K)
poly(2-hydroxyethyl methacrylate)	PHEMA	CH_3	CH_2-CH_2-OH	376	300	140
poly(5-hydroxy-3-oxapentyl methacrylate)		CH_3	$(CH_2-CH_2-O)_2H$			140
poly(8-hydroxy-3,6-dioxaoctyl methacrylate)		CH_3	$(CH_2-CH_2-O)_3H$			140
poly(pivaloyl-2-oxyethyl methacrylate)	PPOEMA	CH_3	$CH_2-CH_2-O-CO-C(CH_3)_3$	313	–	145
poly(methyl methacrylate)	PMMA	CH_3	CH_3	385	283	
poly(ethyl methacrylate)	PEMA	CH_3	CH_2-CH_3	350	273	
poly(n-propyl methacrylate)	PPMA	CH_3	$CH_3-CH_2-CH_3$	332	(285)	95
poly(n-butyl methacrylate)	PBMA	CH_3	$CH_2-CH_2-CH_2-CH_3$	293	–	98
poly(2,2,2-trichloroethyl methacrylate)		CH_3	CH_2-CCl_3	372	273	
poly(2,2,2-trichloro-1-methoxyethyl methacrylate)		CH_3	$CH(OCH_3)-CCl_3$	395	271	
poly(2,2,2-trichloro-1-ethoxyethyl methacrylate)		CH_3	$CH(OCH_2-CH_3)-CCl_3$	399	273	

Table 1 (continued)

Polymer	Abbreviation	R_1	R_2	T_α^a (K)	T_β^a (K)	T_γ^a (K)
poly(2-hydroxyethyl acrylate)	PHEA	H	CH_2-CH_2-OH	282	178	140
poly(methacrylic acid)	PMAAc	CH_3	H		253	
poly(acrylic acid)	PAAc	H	H	381	–	
polymethacrylamide	PMAAm	CH_3	$CO-NH_2^b$			
polyacrylamide	PAAm	H	$CO-NH_2^b$		248	
poly(N-methylmethacrylamide)		CH_3	$CO-NH-CH_3^b$			
poly(N-ethylmethacrylamide)		CH_3	$CO-NH-CH_2-CH_3^b$			
poly(N-n-butylmethacrylamide)		CH_3	$CO-NH-CH_2-CH_2-CH_3^b$			77
poly(N-2-hydroxypropylmethacrylamide)		CH_3	$CO-NH-CH_2-CHOH-CH_3^b$			143

[a] Temperature location of the peak of the loss maximum at frequency 1 Hz: α = main transition from glass to rubberlike state; β = secondary transition related to side-chain rotation; γ = low-temperature transition related to internal motion within the side chain

[b] Whole side chain

Enlargement of the moving group is generally thought to account for a higher activation energy and relaxation temperature. This trend has clearly been proved[1] for a series of poly(alkyl methacrylates), the alkyl groups being CH_3, C_2H_5, n-C_3H_7, and n-C_4H_9. However, the homologs of PHEMA obtained by substituting the 2-hydroxyethyl side group by $(CH_2–CH_2–O)_nH$, where n = 2 or 3, exhibit (Fig. 4) a virtually identical temperature location, activation energy, and intensity (i.e. loss maximum area and the concomitant drop in the storage modulus) of the γ process[65, 146]. As for the constancy of the latter quantity, one can only speculate that the decreasing number of side chains in a volume unit is incidentally compensated by their increasing volume fraction and/or relaxation strength. To assess whether $CH_2–CH_2–O–$ or $CH_2–CH_2–OH$ is significant for the γ process, the dynamic-mechanical response spectrum of poly(pivaloyl-2-oxyethyl methacrylate) (PPOEMA) has been determined[160] (Fig. 5). Since the γ peak is retained at 145 K, one can infer that a substituent attached to the $COO–CH_2–CH_2–O–$ side group does not contribute to the potential energy barrier but affects only the extent of the dissipative mechanism. Furthermore, Fig. 5 shows that stereoregularity has some influence on the height but not on the temperature location of the γ loss peak.

The effect of the side chain bulkiness has been further studied on a series of chloro derivatives of poly(ethyl methacrylate)(PEMA). Though poly(2-chloroethyl methacrylate) exhibits[69] a pronounced peak at $T_\gamma = 117$ K, poly(2,2,2-trichloroethyl methacrylate), poly(2,2,2-trichloro-1-methoxyethyl methacrylate), and poly(2,2,2-trichloro-1-ethoxyethyl methacrylate) do not show (Fig. 6) any low-temperature loss maximum above the liquid nitrogen temperature[157]. However, these three polymers probably display a relaxation process below 77 K as indicated by the decrease in the loss modulus with rising temperature up to 100 K. Their relaxation behavior seems to be similar to that of PEMA rather than of poly(2-chloroethyl methacrylate) which is difficult to explain.

The low-temperature relaxation process of side groups in polymethacrylates is affected not only by their polarity and/or bulkiness but also by steric repulsive forces exerted by substituents of adjacent mers (of the same backbone). If HEMA is copolymerized with a methacrylate having a small side chain, which does not give rise to a γ relaxation above 77 K (Fig. 7), e.g. $COOCH_3$, COOH, $CONH_2$, CN, the T_γ's of the remaining 2-hydroxyethyl groups decrease approximately linearly with the comonomer volume fraction v_H (Fig. 8). Since the potential barrier within the side group cannot be

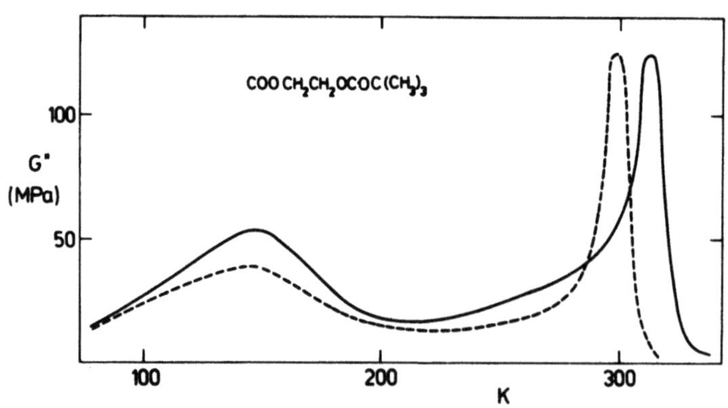

Fig. 5. Temperature dependence of the shear loss modulus of poly(pivaloyl-2-oxyethyl methacrylate): atactic (——), isotactic (– – –)

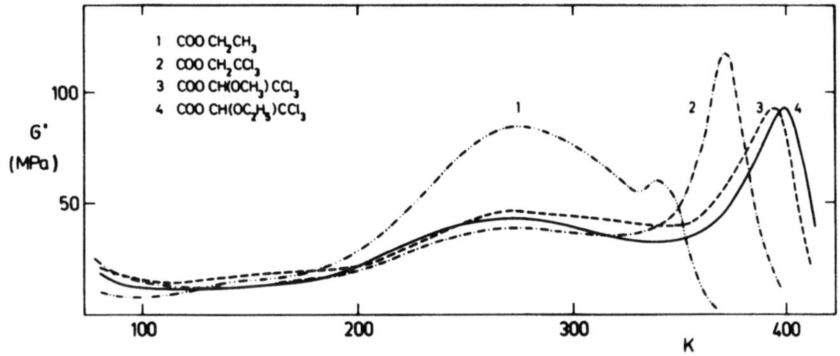

Fig. 6. Temperature dependence of the shear loss modulus of poly(ethyl methacrylate) (*1*), poly(2,2,2-trichloroethyl methacrylate) (*2*), poly(2,2,2-trichloro-1-methoxyethyl methacrylate) (*3*), and poly(2,2,2-trichloro-ethoxyethyl methacrylate) (*4*)

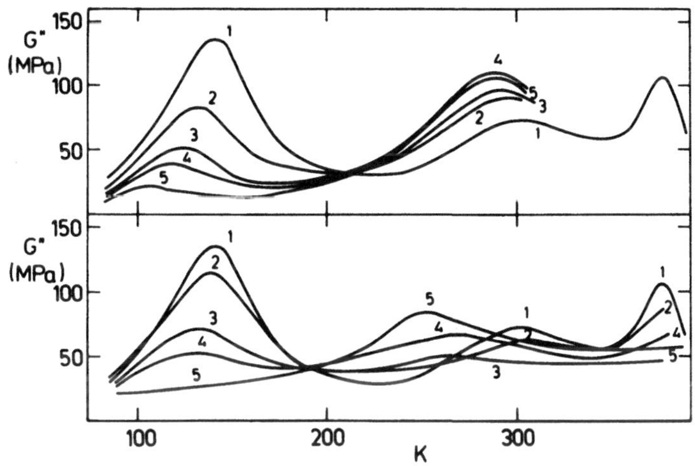

Fig. 7. Effect of the volume fraction of methyl methacrylate (*upper figure*) and of acrylamide (*lower figure*) in copolymers with 2-hydroxyethyl methacrylate on the temperature dependence of the shear loss modulus. MMA: 1 = 0.00; 2 = 0.22; 3 = 0.45; 4 = 0.63; 5 = 0.81. AAm: 1 = 0.00; 2 = 0.19; 3 = 0.51; 4 = 0.79; 5 = 1.00

Fig. 8. Effect of the comonomer volume fraction v_H on the T_γ depression of 2-hydroxyethyl methacrylate Acrylic acid (●), methacrylic acid (○), acrylamide (◑), methacrylamide (◐), acrylonitrile (⊖), methacrylonitrile (◓), and methyl methacrylate (⊖)

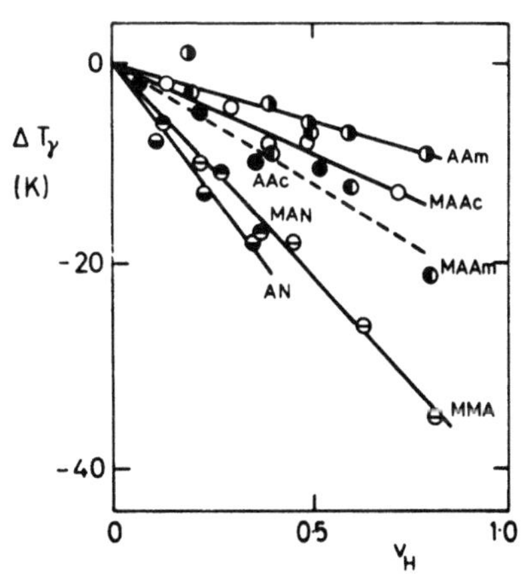

affected by a comonomer, the drop in T_γ can only be understood as a result of a decrease in the contribution to the potential barrier due to adjacent side groups. It is interesting to analyze the effect of various comonomers from the following point of view: methyl methacrylate (MMA) is more efficient than methacrylic acid (MAAc) or acrylic acid (AAc) though MMA has the α-methyl group on the backbone and the methoxycarbonyl group is bulkier than the carboxylic group which evidences a strong countereffect of the polarity of the side group. The superimposing effect of the α-methyl group is not an unambiguous one because AAc or acrylonitrile reduce T_γ somewhat more than MAAc or methacrylonitrile while for the pair acryl amide (AAm) and methacryl amide (MAAm) the relationship is an opposite one. This clearly indicates the high complexity of the combined effects of geometrical constraints and of interactions between various polar side groups.

The α-methyl group on the backbone is known to cause significant differences in the molecular mobility of polymethacrylates and polyacrylates[1]. However, if its concentration is decreased by copolymerizing[127] HEMA with 2-hydroxyethyl acrylate (HEA), the temperature location of the γ loss peak is stable (Fig. 9). This evidences that the energy

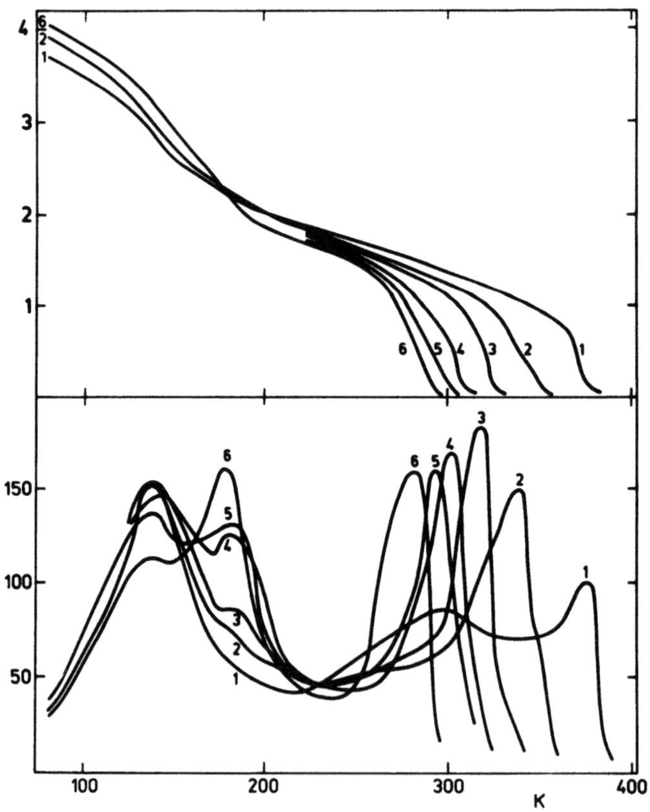

Fig. 9. Effect of the volume fraction of 2-hydroxyethyl methacrylate on the temperature dependence of the moduli G′ and G″ in copolymers with 2-hydroxyethyl acrylate 1 = 1.00; 2 = 0.76; 3 = 0.66; 4 = 0.43; 5 = 0.31; 6 = 0.00

barrier to the side group rotation is independent of the presence of the α-methyl group. On the other hand, there is no plausible explanation of the loss maximum reduction. One of the conceivable reasons may be sought in the alteration of the tacticity (has not been determined), since it has been found[129] that isotactic PHEMA exhibits a much lower γ peak than syndiotactic PHEMA though T_γ remains constant. Similarly, isotactic PPOEMA displays a smaller γ peak than the atactic polymer (Fig. 5).

The individual contributions to the energy barrier ΔU opposing the side-group rotation are difficult to resolve. A substantial part, ΔU^m, originates inside the monomeric unit (mer), similarly to the low-molecular weight compounds. The magnitude of the activation energy of internal rotation in these compounds is primarily determined by the size of the atoms or groups attached to the bond about which internal rotation takes place. At the same time, interactions of permanent and/or induced dipoles can contribute sizeably to the activation energy. The values of ΔU^m for some simple molecules similar to the groups contained in the side chains have been determined[51, 56], e.g. 14 kJ/mol for n-butane, 18.5 kJ/mol for 1-chloro-2-methylpropane, and 20.5 kJ/mol for 1-bromopropane. The values for analogous hydroxy derivatives have not been reported.

Since the T_γ's of the HEMA component in copolymers are affected by both the bulkiness and polarity of the comonomer side chain, it is necessary to take into account a contribution ΔU^c due to the adjacent mers of the same backbone. The extrapolated T_γ of an isolated 2-hydroxyethyl group inside a MMA sequence is lower by 41 K[152] than the T_γ of PHEMA which, according to Eq. (9), represents a drop in the activation energy by about 10 kJ/mol. Analogously, a difference of about 4.8 kJ/mol is obtained from data on copolymers BMA-MMA[152]. This is to illustrate the significance of adjacent substituents for ΔU^c even though it is apparent, on the other hand, that the α-methyl group does not play an appreciable role.

The decrease with temperature (i.e. volume expansion) and increase with pressure of the activation energy, as determined for the rotation of $COOCH_3$, evidences the contribution due to intermolecular interactions, ΔU^e, which might represent[62] up to 20% of ΔU. The ΔU^e's are distributed because of the statistical character of intermolecular forces and may be regarded as the primary cause of the generally observed distribution of the activation energy ΔU. Actually, the values of ΔU of the γ process of PHEMA vary[66] in the range nearly 40 kJ/mol wide which necessitates a distribution of ΔU^c.

If the estimated contributions to the activation energy of the γ relaxation process of PBMA, i.e. $\Delta U^m = 14$ kJ/mol, $\Delta U^c = 5$ kJ/mol, $\Delta U^e = 0$ to 4 kJ/mol, are summed up, the resulting interval 19 to 23 kJ/mol includes the average experimental value[1] 22 kJ/mol. Similarly, the values $\Delta U^m = 20$ kJ/mol, $\Delta U^c = 10$ kJ/mol, $\Delta U^e = 0$ to 7 kJ/mol, somewhat speculatively estimated for PHEMA lead to a ΔU of about 30 to 37 kJ/mol which might be viewed as a reasonable accord with the prediction of Eq. (9).

The extent of the γ relaxation process at 140 K of all copolymers presented above is linearly proportional to the HEMA volume fraction (Fig. 10). This means that the relaxation strength of the 2-hydroxyethyl group does not noticeably depend on the type and concentration of the comonomer with a small side group. In terms of the site model[6, 9], it can be understood that the energy difference between relevant conformers, which is believed to control the relaxation strength, remains constant. This can tentatively be explained by assuming that the energy difference is inherent in the rotation potential within the side chain itself and is not affected, unlike the energy barrier, by the external factors discussed above.

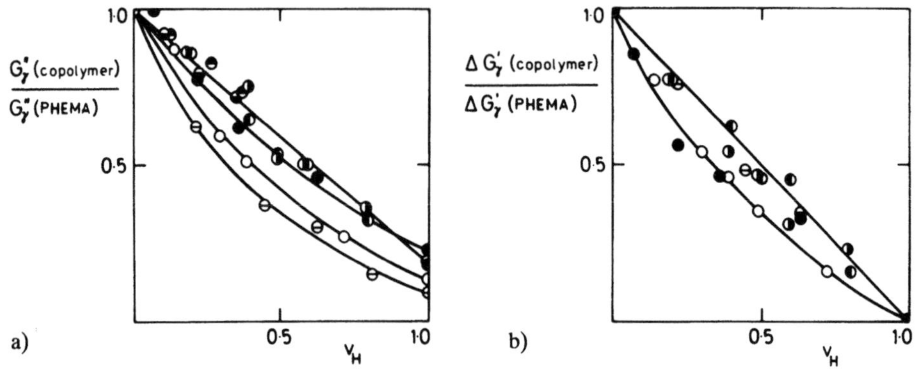

Fiq. 10 a, b. Effect of the comonomer volume fraction v_H on (**a**) the relative height and (**b**) the relative storage modulus decrement of the low-temperature dispersion of 2-hydroxyethyl methacrylate (symbols as in Fig. 8)

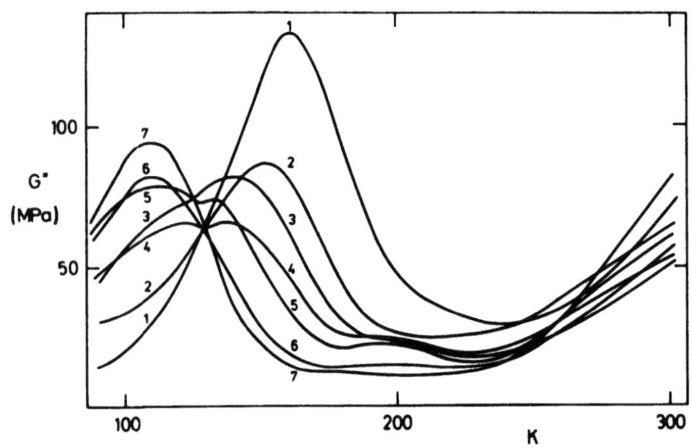

Fig. 11. Effect of the molar fraction of 2-hydroxyethyl methacrylate on the temperature dependence on the shear loss modulus of copolymers with n-butyl methacrylate 1 = 1.00; 2 = 0.80 ; 3 = 0.60; 4 = 0.50; 5 = 0.40; 6 = 0.20; 7 = 0.00

However, if two kinds of side groups, each giving rise (in homopolymers) to a characteristic 2γ process, are mixed in the ratio 1:1 in a random copolymer[159], then the molecular motions affect each other but do not merge (Fig. 11). Two distinct loss maxima existing side by side, though their peaks shift toward each other due to overlapping, provide evidence that the low-temperature relaxations of different side groups retain their identity. (At low concentrations of either component, the corresponding small peak cannot be resolved in the proximity of that of the prevailing component.)

5.2 The β Secondary Relaxation

The partial rotation of side chains[12,13] COOR sets in (Fig. 4) at about $T_\beta = 280$ K, if R = CH_3; C_2H_5; C_3H_7; C_4H_9; consequently. T_β of poly(alkyl methacrylates) is not affected by another alkyl methacrylate comonomer. (Polymethacrylates with longer side chains exhibit the glass transition temperature T_g close to or below T_β so that the secondary transition is overlapped by the glass transition.) Similarly, chloro derivatives of PEMA are characterized by T_β indentical with that of the parent polymer, namely, 273 K (Fig. 6). The higher temperature T_β = 300 K of PHEMA has been attributed[15] to higher polarity and, consequently, to stronger interactions of the side chains. In contrast, PMAAc displays (Fig. 12) an extensive process at T_β = 253 K which is lower by 20 to 47 K than that of the esters just mentioned. Since the interaction of the PMAAc side groups is very strong because of the formation of the hydrogen bonds[163], the temperature depression has to be ascribed to much lower spatial constraints for R = H. Compared with its ordinary esters, PMAAc exhibits the lowest T_β, and the highest T_α (extrapolated value[164] is about 500 K). This clearly indicates the dissimilarity of molecular mechanisms underlying the β and α (glass) transitions in polymethacrylates. Thus, it seems highly unlikely that T_β/T_α would be a constant[165] or would obey a simple empirical rule[166].

Enlargement of the alkyl group accounts for a slight reduction of the extent of the dissipative mechanism because the height and area of the β peak decrease[13]. In conformity with this trend, PMAAc displays much higher energy losses in the β transition than its esters. To explain this, it seems inevitable to assume that the local motion of the main chains plays some role in the β process of PMAAc. It should be noted, however, that in general the extent to which the backbone participates in the side chains motion has not yet been quantified.

There are many examples observed so far that rotation of the COOR side chains in copolymers is affected by substituents on adjacent mers, i.e. by polarity and/or bulkiness of the group R and chiefly by the group in the α-position on the backbone[1]. T_β of copolymers HEMA-MAAc decreases with the volume fraction of MAAc (Fig. 12) while the peak area increases. The alterations of both quantities are quite systematic and continuous[149] which is believed to be an indication of only a slight modification, if any, of the underlying molecular mechanism. On the other hand, the β dispersion of HEMA shifts to higher temperatures in copolymers[136] with increasing content MAAm, which reflects the restrictions to the motions imposed by polar and stiff (planar) amide groups. The absence of α-methyl groups in polyacrylates has as a general consequence lower

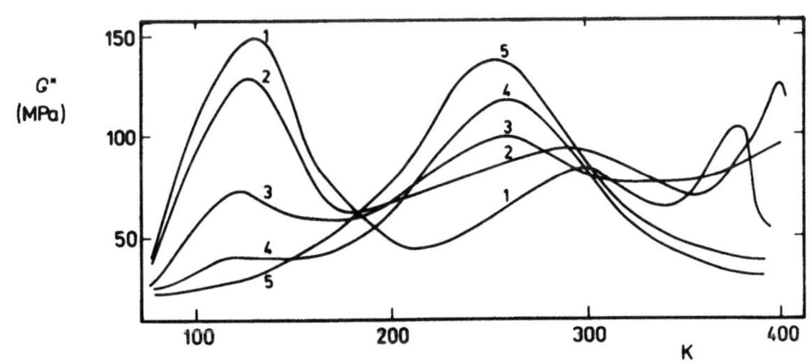

Fig. 12. Effect of the volume fraction of methacrylic acid on the temperature dependence of the shear loss modulus of copolymers with 2-hydroxyethyl methacrylate 1 = 0.00; 2 = 0.14; 3 = 0.39; 4 = 0.72; 5 = 1.00

temperatures of the (glass) α and β transitions; besides, the extent of the β process of polyacrylates is usually much smaller than that of the corresponding polymethacrylates. In the copolymers methyl methacrylate-methyl acrylate[12], methyl methacrylate-acrylonitrile[167] etc., the increasing fraction of acrylate comonomer has caused the β maximum to be reduced and shifted toward lower temperatures. In contrast to the results reported earlier, we have assessed[127] for the copolymer HEMA-HEA that a decrease in the α-methyl group concentration brings about a reduction of the existing β maximum (at 300 K) and the formation of another secondary β' loss maximum at 178 K, typical of PHEA (Fig. 9). However, the enlargement of the β' peak occurs more rapidly with growing acrylate volume fraction. It can be assumed that isolated acrylate mers incorporated into methacrylate sections are not sufficient for producing the β' dispersion but that the formation of acrylate sequences is necessary. The stable temperature position of the β' dispersion provides evidence that the corresponding relaxation motion is controlled by local intramolecular interactions and is independent of the average copolymer composition. The cause of the β' process is to be sought in the onset of the side chain rotation inside the acrylate polyads.

It has also been established for the copolymers HEMA-AAm (Fig. 7) that the temperature and intensity of the secondary dispersion do not change systematically with the copolymer composition. Thus, molecular motions underlying the β or β' dispersions in polymethacrylates or polyacrylates differ from each other. This may be ascribed to a different cooperation of the backbone, but the results obtained so far do not suffice for a more precise interpretation. Likewise, it is difficult to explain[156] that PAAc, in contrast to PMAAc, does not exhibit any secondary relaxation above the liquid nitrogen temperature. It is to be noted that the effect of stereoregularity and diluents[168] on the β relaxation is not easy to estimate because the concomitant decrease in T_α accounts for the overlapping of the β and α transitions.

As in the case of γ relaxation, we can attempt to estimate the individual contributions to the activation energy of the β relaxation process of PHEMA. Since no appropriate model compound has been studied, we have to resort to the value $\Delta U^m = 25$ kJ/mol common to a series of esters[56] (with unbranched alkyls). The value of ΔU^e may be approximated by that determined[62] for PMMA, i.e. 11 kJ/mol. Removal of the methyl group from the backbone causes a drop in T_β by 120 K; according to Eq. (9), it corresponds to $\Delta U^c = 27$ kJ/mol. (The presumed modification of the molecular mechanism due to different participation of the backbone is disregarded.) The sum of the accretions amounts to 63 kJ/mol while the experimentally assessed value[1] is about $\Delta U = 72$ kJ/mol. Nevertheless, this accord may be regarded as reasonable because the actual values of ΔU^m and ΔU^c are obviously somewhat higher.

5.3 Effect of Diluents and the Diluent-Induced Secondary Relaxation

Most conspicuous modifications of the dynamic mechanical response spectra of PHEMA and related polymers are brought about by incorporation of low-molecular weight compounds (Fig. 13). Along with alterations of parameters (temperature, height, shape) of the peaks characteristic of a dry polymer, usually a new diluent peak appears. (The relaxation patterns of various polymethacrylates are not modified by diluents in a unique way but several modes can be distinguished as mentioned before.) A remarkable feature

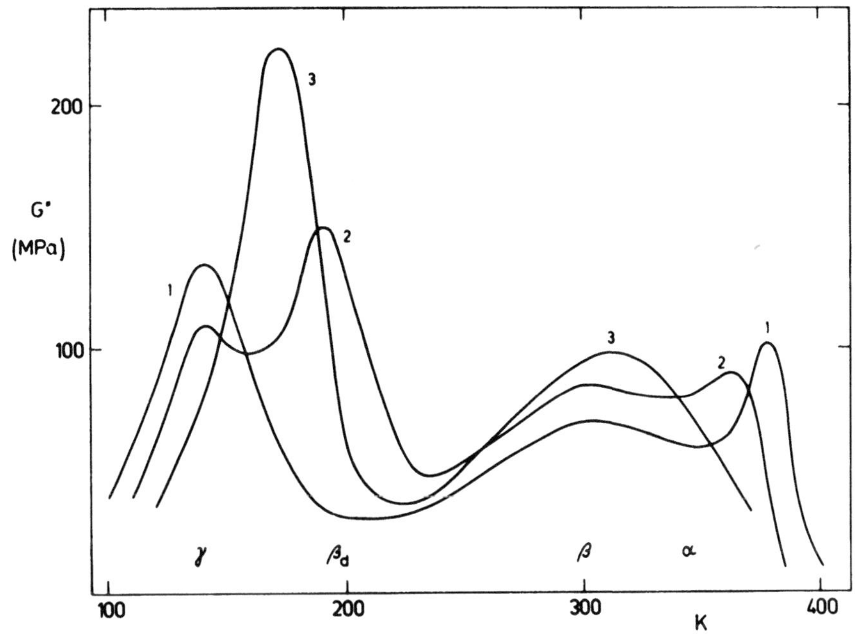

Fig. 13. Effect of the volume fraction of water on the temperature dependence of the shear loss modulus of poly(2-hydroxyethyl methacrylate) 1 = 0.00; 2 = 0.02; 3 = 0.09

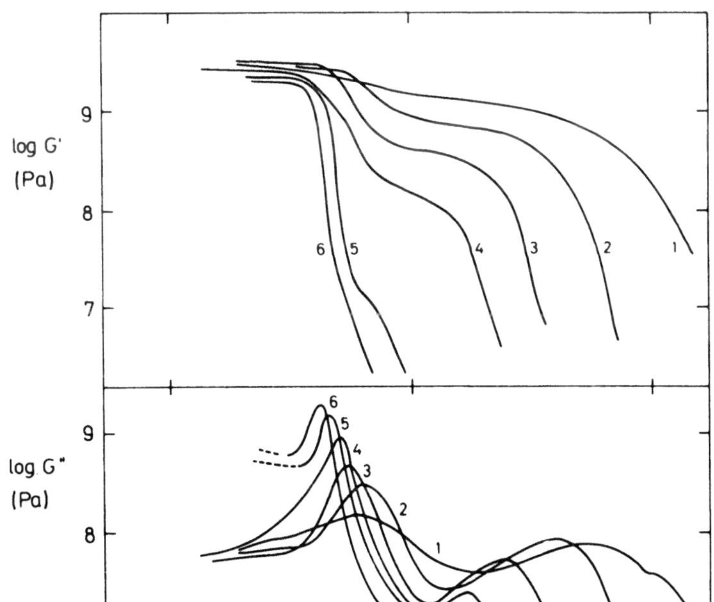

Fig. 14. Effect of the volume fraction of ethylene glycol on the temperature dependence of the shear storage and loss moduli of poly(2-hydroxyethyl methacrylate) 1 = 0.15; 2 = 0.33; 3 = 0.47; 4 = 0.58; 5 = 0.76; 6 = 0.84

of such diluent dispersions is that they are induced by traces of diluents[1, 15, 148], or residues of monomers and oligomers[100] etc. A typical example is represented by PHEMA (Fig. 13) whose γ relaxation is reduced and eventually vanishes with increasing diluent concentration; simultaneously, a diluent β_d relaxation appears at about 200 K which is proportionally intensified (Fig. 14) and shifted (Fig. 15) to lower temperatures. A quite analogous behavior has been observed with poly(propyl methacrylate)[151], poly(N-2-hydroxypropyl methacrylamide)[155], and poly(N-butyl methacrylamide)[155]. The conclusion has been forwarded[136, 146, 148, 151] that the interaction of a diluent molecule with the hydroxy or also with the ester or amide group of the side chain leads to a transformation of the γ process into the β_d process.

The second type of the diluent-induced alterations in the relaxation pattern is characteristic of e.g. PMAAc[149] (Fig. 16), PAAm[136] (Fig. 17), and PHEA[127] (Fig. 18). Incorporation of water (no other diluent has been used) accounts for a continuous decrease in the temperature and increase in the area of the existing β or β' loss peak. Since no new peak is produced, it seems reasonable to infer that the mobility of diluent molecules sets in along with the partial rotation of the COOR side chains. However, it is possible to

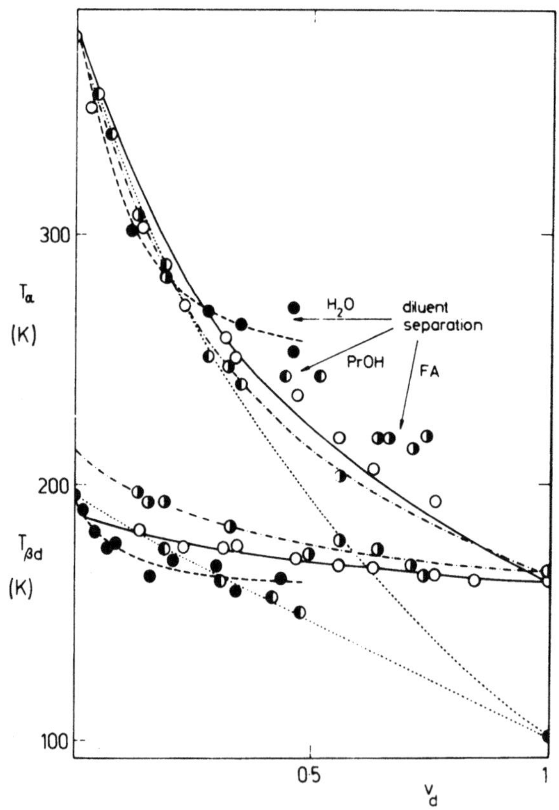

Fig. 15. Effect of the diluent volume fraction on the temperature of the T_α and $T_{\beta d}$ dispersions of poly(2-hydroxyethyl methacrylate) Ethylene glycol (—, ○); formamide (-.-, ◑); water (----, ●); propanol (. . ., ◐). The curves have been calculated according to Eq. (13), for the system PHEMA-water drawn through experimental points

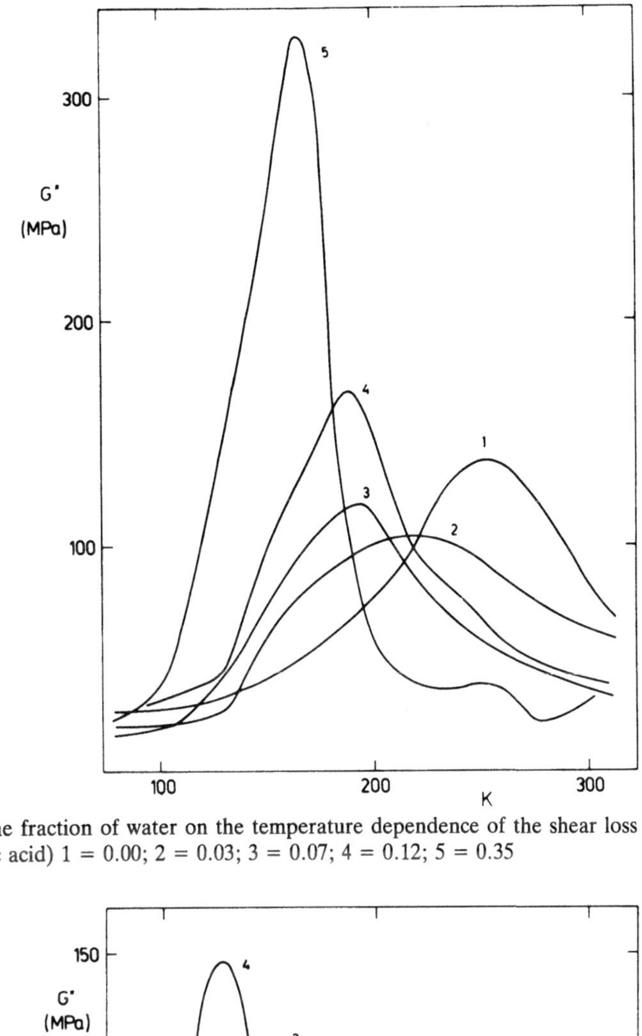

Fig. 16. Effect of the volume fraction of water on the temperature dependence of the shear loss modulus of poly(methacrylic acid) 1 = 0.00; 2 = 0.03; 3 = 0.07; 4 = 0.12; 5 = 0.35

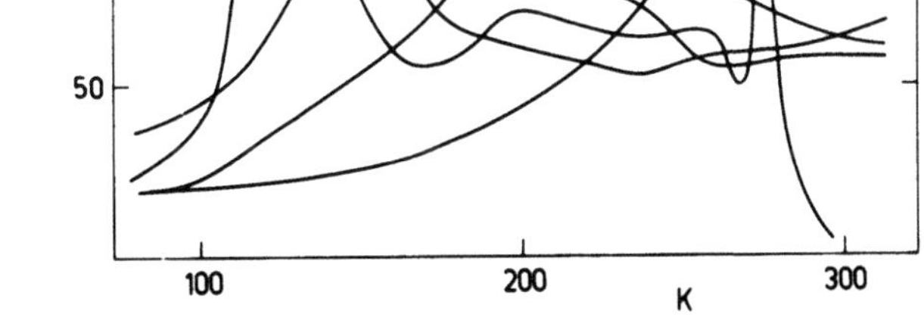

Fig. 17. Effect of the volume fraction of water on the temperature dependence of the shear loss modulus of polyacrylamide 1 = 0.00; 2 = 0.06; 3 = 0.13; 4 = 0.36

speculate that a new diluent peak is formed which is superimposed upon the existing peak displayed by the dry polymer.

Another group encompasses polymethacrylates that do not exhibit any loss maximum (in the dry state) between 77 and 300 K, e.g. PAAc[156] (Fig. 19), PMAAm[136] (Fig. 20),

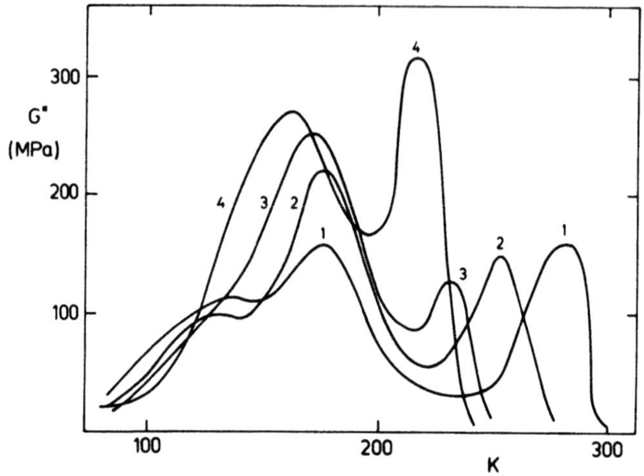

Fig. 18. Effect of the volume fraction of water on the temperature dependence of the shear loss modulus of poly(2-hydroxyethyl acrylate) 1 = 0.00; 2 = 0.07; 3 = 0.16; 4 = 0.46

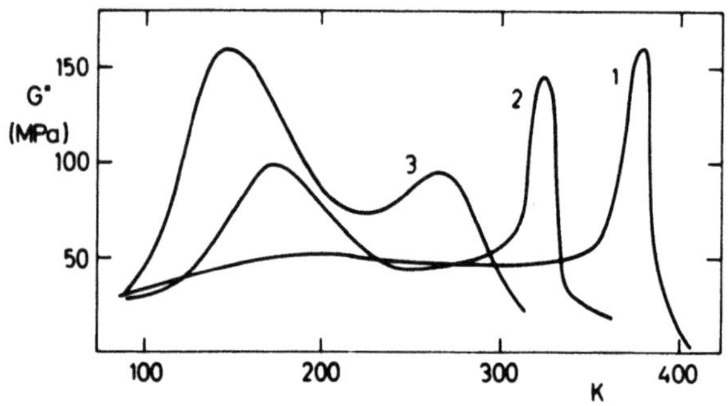

Fig. 19. Effect of the volume fraction of water on the temperature dependence of the shear loss modulus of poly-(acrylic acid) 1 = traces; 2 = 0.06; 3 = 0.33

poly(N-ethylmethacrylamide)[155] etc., but are characterized by a loss peak situated between 120 and 200 K if water (and possibly other diluents) is incorporated. The size of such a peak is proportional to the water content while the loss peak temperature is decreased (PAAc) or remains constant (PMAAm). As mentioned above, the occurrence of such loss peaks is most propable due to some motion of the diluent molecules or of complexes consisting of a diluent molecule (or molecules) and a side group of the polymer (though the side group itself does not give rise to a relaxation).

A common feature of all polymethacrylates, whatever group they belong to, is that the extent of the diluent-induced relaxation process is proportional to the diluent content. This relation has been studied[148] in great detail on PHEMA swollen with ethylene glycol (Fig. 14), formamide, propanol, and water. Apart from the increase in the loss peak magnitude, also the concomitant decrement of the storage modulus, $\Delta G'_{\beta d} = G'_g - G'_b$, (Fig. 21) rises at the expense of the decrement belonging to the main transition, i.e. $\Delta G'_a = G'_b - G_e$. It is known that $\Delta G'$ associated with a dispersion in glassy polymers is much less than one order of magnitude whereas in the transition region from the glassy to the rubberlike state, the modulus drops by three to four orders of magnitude. As can be

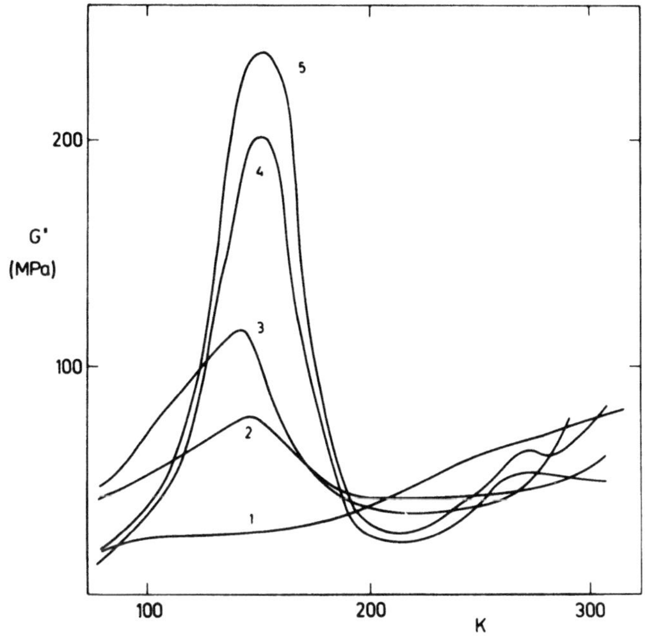

Fig. 20. Effect of the volume fraction of water on the temperature dependence of the shear loss modulus of polymethacrylamide 1 = 0.00; 2 = 0.05; 3 = 0.12; 4 = 0.35; 5 = 0.40

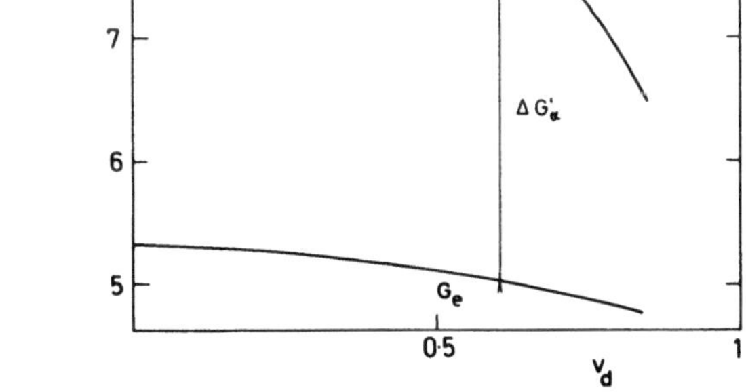

Fig. 21. Effect of the volume fraction v_d of ethylene glycol on the decrease in the shear storage modulus in the transition from the glassy to rubberlike state. G'_g is the modulus of the glassy state (140 K), G'_b represents the boundary between the β_d and α dispersions (cf. Fig. 14), G_e is the modulus in the rubberlike state

seen from Fig. 21, $\Delta G'_{\beta d}$ rises with ethylene glycol content and for $v_d > 0.5$ exceeds one logarithmic decade and becomes commensurable with $\Delta G'_\alpha$. All the data provide evidence that the intensity of the β_d dispersion increases with the diluent content not only in the region of low v_d, in which there is a rise in the number of the side chains interacting with the diluent molecules and participating in the β_d process, but also in the region of higher v_d, in which the number of these side chains per unit volume of the system decreases because of dilution. It can be concluded[148], therefore, that at $T_{\beta d}$ the motion of diluent molecules sets in, along with that of the side chains, and contributes largely to the intensity of the β_d dispersion. The role of diluent molecules is even more important in polymers without any secondary relaxation which could be transformed or intensified. Data collected to date for these polymers are insufficient for estimating the participation of the side and/or main chains and elucidating the underlying mechanisms.

As is obvious from Figs. 13 and 14, T_α and $T_{\beta d}$ of swollen polymers decrease with increasing diluent volume fraction v_d. It has been observed[148] for the system PHEMA – diluent that both temperatures approach $T_{\alpha d}$ of the diluent when v_d approaches unity (Fig. 15). The concentration dependence of the T'_αs can be described by an equation derived[169] by assuming additivity of the free volumes of the components:

$$T_\alpha = T_{\alpha p} + (k_\alpha T_{\alpha d} - T_{\alpha p})v_d/[1 + (k_\alpha - 1)v_d] \, , \tag{13}$$

where $T_{\alpha p}$ is the main transition temperature of dry PHEMA (read off as the peak of the loss maximum) and k_α a constant corresponding to the ratio of the expansion coefficients of the free volumes of diluent and polymer. For the system PHEMA – ethylene glycol, which has remained homogeneous (upon cooling down) over the whole range of concentrations measured, k_α approximately equals (Table 2) the free volumes ratio determined independently[170]. In the concentration range in which no signs of a phase separation (as crystals or glass) of formamide ($v_d < 0.64$) or propanol ($v_d < 0.4$) have been observed, Eq. (13) also provides reasonable values of k_α. Water separation appears in the cooled-down systems already at about $v_d = 0.25$ which prevents T_α from decreasing below 263 K.

Equation (13) also allows to describe the $T_{\beta d}$ vs. v_d dependence when T_α is replaced by $T_{\beta d}$ and $T_{\alpha p}$ by $T^0_{\beta d}$, i.e. by a hypothetical temperature of the β_d process at $v_d = 0$. The resulting constant $k_{\beta d}$ is a mere characteristic of the curvature of this dependence and cannot be assigned the original physical meaning because there is no β_d process and no corresponding rise in the expansion coefficient for dry PHEMA. It can be seen from Table 2 that $T^0_{\beta d}$ and $k_{\beta d}$ are dependent on the type of diluent which evidences the role of polymer-diluent interactions. It should be emphasized that $T^0_{\beta d}$ is not a characteristic of the polymer itself but of a pair polymer-diluent. This result has also been corroborated for the opposite type of systems, i.e. for different polymers swollen with the same diluent, namely, water. The temperature location of the water-induced β_d loss maxima and its variation with the water concentration clearly depend on the polymer composition (cf. Figs. 13, 16–20).

Table 2. Coefficients k_α and $k_{\beta d}$ of Eq. (13) and temperature $T^0_{\beta d}$ for poly(2-hydroxyethyl methacrylate)-diluent systems

Diluent	k_α	$k_{\beta d}$	$T^0_{\beta d}$(K)
Ethylene glycol	3.0	2.6	187
Formamide	3.6	3.6	213
Propanol	(2.1)	(1.1)	(193)
Water	–	–	192

Since no exception has been observed to date, it seems that $T_{\beta d}$ is always higher than T_g of the incorporated diluent. This holds not only for PHEMA swollen with various diluents, but also for PMAAc[149], PAAc[156], PMAAm[136], PAAm[136], poly(N-substituted methacrylamides)[155] swollen with water. The lowest temperature of the β_d dispersion induced by water has been observed for PAAm (Fig. 17), namely 135 K at $v_d = 0.36$ (a loss peak near 273 K indicates partial separation of water during cooling), which lies within the range of the reported[171] values of the glass transition temperature of water, i.e. 127 to 140 K. Taking into account that $T_{\beta d} > T_{\alpha d}$ and that Eq. (13) can describe the $T_{\beta d}$ vs. v_d dependence, it seems likely that $T_{\beta d}$ is codeterminated by the free volume of the polymer-diluent system in the glassy state.

Now the question can be raised which effect on the molecular mobility would have a diluent with $T_{\alpha d}$ above the temperatures of ordinary β_d dispersions. On the basis of

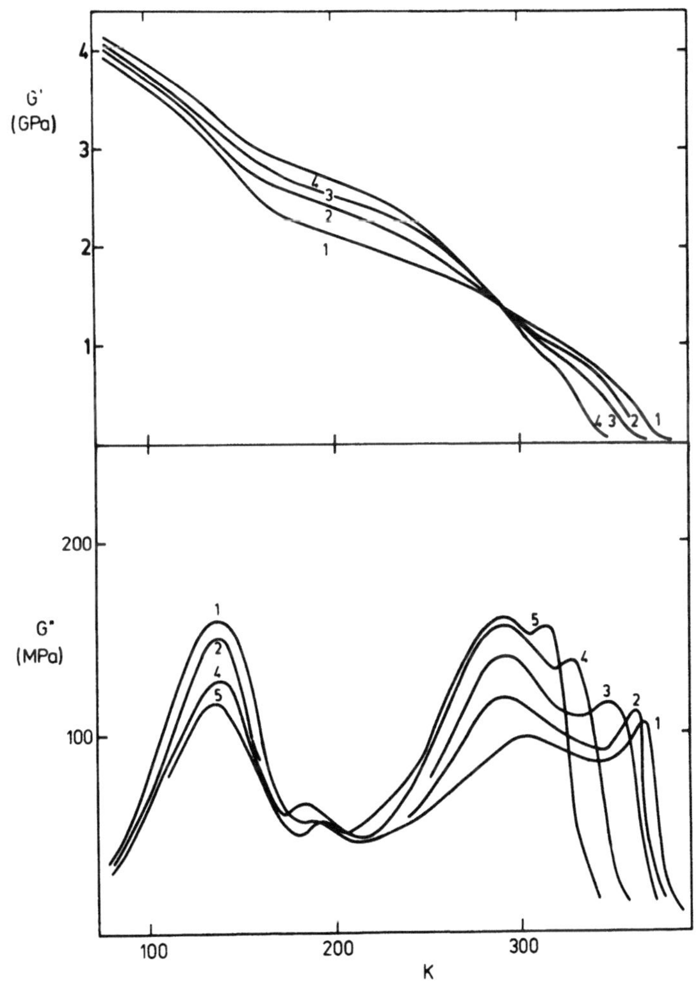

Fig. 22. Effect of the weight fraction of urea on the temperature dependence of the storage and loss moduli of poly(2-hydroxyethyl methacrylate) 1 = 0.00; 2 = 0.021; 3 = 0.039; 4 = 0.068; 5 = 0.115

earlier results[172] urea, whose melting point is $T_m = 406$ K[173] has been used for this purpose[159]. According to the empirical rule $0.5\ T_m < T_g < 0.67\ T_m$, its T_g value can be expected to lie between 203 and 271 K. (Diffraction patterns have shown that the samples contain no crystalline urea up to an urea weight fraction $w_u \leq 0.115$). With increasing urea content (Fig. 22), T_α decreases, T_β drops by about 10 K whereas T_γ remains constant; simultaneously, α and β maxima rise, partly due to their overlapping, the γ maximum, on the other hand, is reduced. The samples show only a residual β_d hump whose magnitude is independent of the urea content (in all probability[159], it is induced by traces of other diluents). Thus, urea affects the PHEMA relaxations very much like other diluents do, but it does not give rise to a diluent dispersion below 273 K. The absence of the β_d dispersion is atypical because all the other diluents used to date have produced a loss peak with a height of about 100 MPA at $v_d = 0.1$.

The effect of increasing urea content is also reflected in the temperature dependence of the storage modulus G′: the drop of G′ within the glass transition becomes more pronounced while the decrease in the temperature interval $T_\gamma - T_\beta$ diminishes and the level of the storage modulus is raised. All the determined curves intersect at a temperature close to T_β. This indicates (Fig. 22) that the molecular mobility is suppressed at $T < T_\beta$ and enhanced at $T > T_\beta$. Thus, we can infer that a certain motion of the urea molecules is initiated at T_β, most probably along with that of the oxycarbonyl groups. The qualitatively different effect of urea may be caused by its high (hypothetical) T_g, high potential barrier of internal rotation[174] (79 kJ/mol), and strong interaction with PHEMA leading to a decrease in the free volume in the glassy state. As visualized in Fig. 23, the density of the system is higher than that calculated from the volume additivity of the components, even though the density of the crystalline urea has been used.

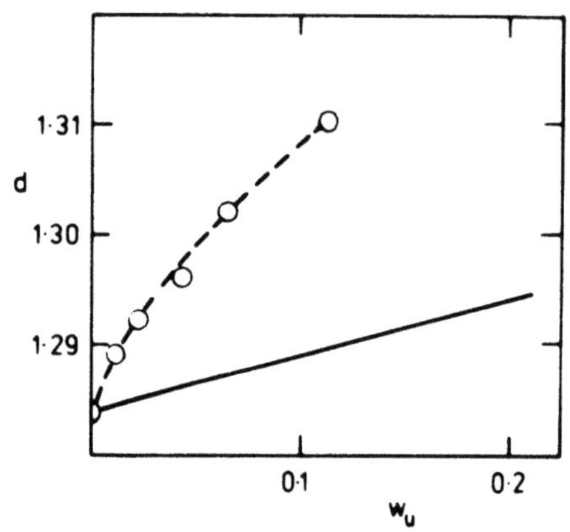

Fig. 23. Density (in Mg/m^3) of the system poly(2-hydroxyethyl methacrylate)-urea as a function of the weight fraction w_u of urea

6 Conclusion

Experimental data collected so far show that most secondary relaxations (transitions, dispersions) in glassy polymers are a consequence of the conformational isomerization of short sections of main or side chains and that their kinetics may be satisfactorily described by means of the site model in which stable conformations are separated by a potential barrier. These relaxations can be effectively studied by employing dynamic methods based on the interference of molecular motions with an oscillating force field. Dynamic mechanical measurements represent the most universal method because they are able to detect the motions of almost all types of structural units (presumably with the exception of the rotation of methyl groups). Because of the kinetic character of secondary transitions, data on their location on the temperature scale must be supplemented by the frequency of measurement. However, the models forwarded so far have not been worked out to such a degree as to give a satisfactory explanation of the mechanism of energy losses on the molecular level and an adequate description of the extent of relaxation processes in dynamic mechanical experiments. At present, verification and refining of the existing theories are difficult, primarily because of the lack of a complete set of pertinent dynamic mechanical data.

When comparing the results obtained (advantageously by using different methods) for a systematically selected series of polymers, it is usually possible to determine, quite reliably, structural units whose motion underlies a certain type of relaxation. On the other hand, it is much more difficult to answer questions regarding the type of motion of an identified motional unit and the mechanism of energy storage and transfer. Molecular motions giving rise to secondary relaxations in glassy polymers were earlier[12, 13] tentatively divided into four types which are documented in this review by further data:

- local main-chain motions assuming rotations of four to six groups about colinear bonds at the ends of the sequences;
- side-chain rotations about the bond linking it to the main chain (possibly with some cooperation of the backbone);
- internal motions within the side chain
- molecular motions of or affected by molecules of a diluent.

The latter group of molecular motions is the least studied up to now; this review attempts to give a categorization of these motions. The classification used is based on the assumption that the motions of diluent molecules are either independent of the molecular mobility of the host glassy polymer or are associated with it in some way. The diluent-induced relaxations, which are assigned to the motions of complex motional units consisting of molecule(s) of the diluent and of a group in monomeric unit of the polymer, are distinguished according to the type of motion exhibited by this group in the absence of the diluent. Although in a particular polymer only a certain type of diluent-induced relaxation motion is assumed, one may expect considerable differences in the molecular mechanism in different polymers:

- motions of or taking place within molecules of the diluent dissolved in the polymer;
- motions of the diluent molecules associated with
  - local main-chain motions;
  - side-chain rotations
  - internal motions within side chains

With the exception of local main-chain motions, the above-mentioned types of molecular motions have been investigated on a series of hydrophilic polymethacrylates and polyacrylates by means of dynamic mechanical measurements carried out with a torsional pendulum. For this purpose, the constitution of polymethacrylates was systematically altered and correlated with the dynamic mechanical response spectra. It was established for a series of copolymers of poly(2-hydroxyethyl methacrylate) that the temperature of the γ relaxation (140 K; 1 Hz), assigned to the motion of 2-hydroxyethyl side groups, was independent of the concentration of α-methyl groups on the main chain but distinctly affected by comonomers, specifically by their polarity and volume. The results were interpreted in terms of the potential barriers and steric hindrances impeding side-chain motions. Evidence collected so far shows that the activation energy may be regarded as consisting of several contributions arising

(a) inside monomeric units,
(b) from the effect of the bulkiness and polarity of the side chains of adjacent mers (in the same backbone), and
(c) from intermolecular interactions.

The individual contributions were estimated (with the partial use of data reported in literature) and their sum found to be in reasonable agreement with experimental data on the activation energy. The relaxation strength of the 2-hydroxyethyl group is independent of the type and concentration of comonomers. This can be explained by assuming that the energy difference between the relevant conformers, which controls the magnitude of dissipation of mechanical energy, is an inherent property of the rotational potential inside the side chains and is not affected by external factors.

The temperature position of the secondary β relaxation (about 290 K; 1 Hz), generally attributed to partial rotations of the side chains COOR, is only slightly affected by the polarity and volume of the substituent R but decreases markedly (by 120 K) on removal of the α-methyl group on the main chain. The experimental data obtained contradict the assumption that there is a certain relationship between this temperature and the glass transition temperature. Nevertheless, we can infer that the pertinent molecular mechanism in polymethacrylates differs from that in polyacrylates, probably due to the different participation of the main chains. The values of the individual contributions to the activation energy were estimated by employing a procedure similar to that used in the γ relaxation process, and their sum was found to agree approximately with the experimental values.

Very conspicuous and diversified modifications of the dynamic mechanical response spectra of polymethacrylates are produced by incorporated diluents. In the case of poly(2-hydroxyethyl methacrylate), the γ relaxation process is reduced and eventually disappears with increasing concentration of the diluent, while a new β_d relaxation process (at very low concentrations of diluents it appears near 200 K) is proportionately intensified and shifted to lower temperatures. Thus, the interaction of diluent molecules with the hydroxy groups (but in other cases also with ester or amide groups) of side chains leads to the transformation of the relaxation processes. Diluent molecules, the motion of which sets in simultaneously with that of the side chains, contribute considerably to the β_d relaxation process. Both the temperature and intensity of diluent-induced relaxations are a function not only of the concentration and glass transition temperature of the diluent but also of the constitution of the polymer. Since no exception could be observed, it seems likely that the temperature of these relaxations is always higher than that of the glass transition of the incorporated diluent. Urea used as a model diluent with a high glass transition temperature displayed a qualitatively different influence on the relaxation behavior of poly(2-hydroxyethyl methacrylate), which could not be satisfactorily explained. Water molecules incorporated into poly(2-hydroxyethyl acrylate), polyacrylamide and polymethacrylic acid induce an enlargement of the existing secondary loss maximum and its displacement toward lower temperatures. Since no new relaxation appears in the response spectrum, it seems probable that the motion of water molecules is associated with the rotation of the side chains. Polymethacrylamide, poly(N-ethylmethacrylamide), polyacrylic acid etc., which lack a secondary dispersion in the range 77 K to T_g, exhibit a pronounced loss maximum after the incorporation of water. It is likely to be brought about by the motion either of diluent molecules or of complexes composed of diluent molecule(s) and side groups of the polymer.

By using dynamic mechanical and some further methods, a large amount of experimental data on the molecular mobility of various polymers and polymer systems has been collected during the last twenty years. In general, extensive research in this area has so far been undertaken. However, there are comparatively few examples of a systematic study of particular polymers or problems of molecular mobility (see e.g.[13, 62, 79, 99]). Thus, a detailed understanding of molecular mechanisms advances rather slowly. As has been pointed out, even a more qualitative interpretation of apparently simple relaxation motions in amorphous polymers is often questionable. Incomplete experimental data, which are provided e.g. by routine isochronous dynamic mechanical measurements, necessarily lead to insufficiently founded or speculative conclusions on the molecular mechanism of the relaxation phenomena observed. It may be difficult to compare results obtained by various authors because data on the frequency of measurements are often lacking, the preparation and thermal history of samples are described insufficiently and traces of unremoved low-molecular weight compounds (of all types) modify the temperature and intensity of relaxations characteristic of the dry polymer, giving sometimes rise to further relaxations. Progress in this field may be achieved by comprehensive experiments and a combination of (complementary) results obtained by several methods, along with the development of the theory.

In spite of all these shortcomings, dynamic mechanical response spectra within a wide temperature range are frequently utilized characteristics of solid-state polymers because they provide direct information on the molecular mobility[1, 5]. The modulus and damping as such are important material constants or functions, if e.g. the reinforcing effect of fillers[3, 175], damping of mechanical vibrations[5] or sonic waves[176] are studied. The extent of relaxation processes in glassy polymers or polymer composites expresses their ability to dissipate mechanical energy, and consequently codetermines the impact resistance[176–180], fracture energy[181] and fatigue crack propagation[182, 183]. Though the molecular mechanism of some relaxations in the dynamic mechanical response spectrum remains unclear, a comparison of the spectra allows conclusions to be drawn on the composition and structure of polymers and changes, due to chemical or physical processes. Up to now, relaxation spectroscopy has been successfully applied to the elucidation of various effects and phenomena reflected in molecular mobility, such as the presence of low-molecular weight compounds[1, 100], their phase separation[15, 148] and plasticizing or anti-plasticizing effect[159, 184]; miscibility of the polymers[176, 185]; interaction of phases in polymeric composites[3, 175, 186]; morphology of crystalline polymers and chain orientation[187–189]; effect of counterions and water in ionene polymers[190, 191]; composition of copolymers[1, 11, 15, 127, 136]; chemical reactions in solid polymer such as cross-linking[192], thermal degradation[193], and photodegradation[194].

7 References

1. McCrum, N. G., Read, B. E., Williams, G.: Anelastic and dielectric effects in polymeric solids. New York: Wiley 1967
2. Haward, R. H. (ed.): The physics of glassy polymers. London: Applied Science Publishers 1973
3. Nielsen, L. E.: Mechanical properties of polymers and composites. New York: M. Dekker 1974
4. Hedvig, P.: Dielectric spectroscopy of polymers. Bristol: A. Hilger 1977
5. Murayama, T.: Dynamic mechanical analysis of polymeric material. Amsterdam: Elsevier 1978
6. Ward, I. M.: Mechanical properties of solid polymers. London: Wiley-Interscience 1971
7. Hill, N. E. et al.: Dielectric properties and molecular behavior. London: Van Nostrand Reinhold 1969
8. Read, B. E., Dean, G. D.: The determination of dynamic properties of polymers and composites. Bristol: A. Hilger 1978
9. Staverman. A. J.: Proceedings of the 5th International Congres on Theology. Onogi, S. (Ed.), Vol. 1, Tokyo: University of Tokyo Press 1969
10. Hedvig, P.: Radiation chemistry of macromolecules. Dole, M. (ed.), Vol. 1. New York: Academic Press 1972
11. Heijboer, J.: Commun. No. 435, Central Laboratorium TNO, Delft 1972
12. Heijboer, J.: An. N. Y. Acad. Sci. *279,* 104 (1976)
13. Heijboer, J.: Int. J. Polym. Mater. *6,* 11 (1977)
14. McCullough, R. L.: J. Macromol. Sci., Phys. *B 9,* 97 (1974)
15. Janáček, J., Kolařík, J.: Int. J. Polym. Mater. *5,* 71 (1976)
16. Haldon, R. A., Schell, W. J., Simha, R.: J. Macromol. Sci., Phys. *B 1,* 759 (1967)
17. Haldon, R. A., Simha, R.: Macromolecules *1,* 340 (1968)
18. Wilson, P. S., Lee, S., Boyer, R. F.: Macromolecules *6,* 914 (1973)
19. Heydemann, P., Guicking, H. D.: Kolloid Z., Z. Polym. *193,* 16 (1963)
20. Quach, A., Simha, R.: J. Appl. Phys. *42,* 4592 (1971)
21. Koppelmann, J., Gielessen, J.: Kolloid Z. *175,* 97 (1961)
22. Saito, S. et al.: J. Polym. Sci. *A-2,* 6, 1297 (1968)
23. O'Reilly, J. M., Karasz, F. E.: J. Polym. Sci. *C 14,* 49 (1966)
24. Struik, L. C. E.: Physical ageing in amorphous polymers and other materials. Amsterdam: Elsevier 1978
25. Anton, A.: J. Appl. Polym. Sci. *12,* 2117 (1968)
26. Hannon, M. J., Koenig, J. L.: J. Polym. Sci. *A-2,* 7, 1985 (1969)
27. Lunn, A. C., Yannas, I. V.: J. Polym. Sci., Polym. Phys. Ed. *10,* 2189 (1972)
28. Hedvig, P.: J. Polym. Sci. *C 42,* 1271 (1973)
29. Zielinski, M., Swiderski, T., Kryszewski, M.: Polymer *19,* 883 (1978)
30. Sessler, G. M. (ed.): Electrets. Berlin: Springer 1980
31. Lacabanne, C., Chatain, D. G., Monpagens, J. C.: J. Macromol. Sci., Phys. *B 13,* 537 (1977)
32. Berticat, P. et al.: J. Macromol. Sci., Phys. *B 15,* 549 (1978)
33. Charlesby, A., Patridge, R. H.: Proc. Roy. Soc. *A 271,* 170 (1963)
34. Partridge, R. H.: J. Polym. Sci. *A 3,* 2817 (1965)
35. Hassan, A. M.: CRC Crit. Rev. Macromol. Sci. *1,* 399 (1972)
36. Allen, G.: Amorphous Materials, Douglas, R. W., Ellis, B. (ed.). London: Wiley 1972
37. Schaefer, J., Stejskal, E. O., Buchdahl, R.: Macromolecules *10,* 384 (1977)
38. Kumler, P. L., Boyer, R. F.: Macromolecules *9,* 903 (1976)
39. Boyer, R. F., Kumler, P. L.: Macromolecules *10,* 461 (1977)
40. Törmälä, P.: J. Macromol. Sci., Rev. Macromol. Chem. *C 17,* 297 (1979)
41. Bullock, A. T., Cameron, C. G., Smith, P. M.: Eur. Polym. J. *11,* 617 (1978)
42. Sung, C. S. P., Lamarre, L., Tse, M. K.: Macromolecules *12,* 666 (1979)
43. Eisenbach, C. D.: Makromol. Chem. *179,* 2489 (1978); *180,* 565 (1979)
44. Jarry, J. P., Monnerie, L.: Macromolecules *12,* 927 (1979)
45. Frank, C. W.: Macromolecules *8,* 305 (1975)

46. Somersall, A. C., Dan, E., Guillet, J. E.: Macromolecules *7,* 233 (1974)
47. Davies, M.: Transfer and storage of energy by molecules. Burnett, G. M., North, A. M. (eds.), Vol. 3. London: Wiley-Interscience 1970
48. Heijboer, J.: Polym. Eng. Sci. *19,* 664 (1979)
49. Neumann, R. M., Senich, G. A., MacKnight, W. J.: Polym. Eng. Sci. *18,* 624 (1978)
50. Krausz, A. S., Eyring, H.: Deformation kinetics. New York: Wiley-Interscience 1975
51. Lamb, J.: Physical acoustics, IIA, Mason, W. P. (ed.). New York: Academic Press 1965
52. Orville-Thomas, W. J., Wyn-Jones, E.: Transfer and storage of energy by molecules. Burnett, G. M., North, A. M. (eds.), Vol. 2. London: Wiley-Interscience 1969
53. Pethrick, R. A., Wyn-Jones, E.: Q. Rev. Chem. Soc. *23,* 301 (1969)
54. Wyn-Jones, E., Pethrick, R. A.: Topics in stereochemistry. V, Eliel, E. L., Allinger, N. L. (eds.), New York: Wiley-Interscience 1970
55. Pethrick, R. A.: Sci. Progr. Oxf. *58,* 563 (1970)
56. Blandamer, M. J.: Introduction to chemical ultrasonics. London: Academic Press 1973
57. Bowden, P. B.: Polymer *9,* 449 (1968)
58. Holliday, L., White, J. W.: Pure Appl. Chem. *26,* 545 (1971)
59. Ward, I. M. (ed.): Structure and properties of oriented polymers. London: Applied Science Publ. 1975
60. Joseph, S. H.: J. Polym. Sci., Polym. Phys. Ed. *16,* 1071 (1978)
61. Boyd, R. H., Breitling, S. M.: Macromolecules *7,* 855 (1974)
62. Koppelmann, J.: Progr. Colloid Polym. Sci. *66,* 235 (1969)
63. Sasabe, H., Saito, S.: J. Polym. Sci. A-2, *6,* 1401 (1968)
64. Read, B. E., Williams, G.: Trans. Faraday Soc. *57,* 1979 (1961)
65. Kolařík, J., Ilavský, M.: Collect. Czech Chem. Commun. *35,* 2286 (1970)
66. Kolařík, J.: Colect. Czech. Chem. Commun. *36,* 2049 (1971)
67. Schatzki, T. F.: Polym. Prepr., Am. Chem. Soc. *17,* 71 (1976)
68. Ferry, J. D.: Viscoelastic properties of polymers. New York: Wiley 1970
69. Waterman, H. A. et al.: Amorphous Materials, Ellis, B. (ed.). London: Wiley 1972
70. Heijboer, J.: Molecular basis of transitions and relaxations, Meier, D. J. (ed.). Midland macromol. monographs, Vol. 4, 1978
71. Koppelmann, J., Leder, H., Royer, F.: Colloid Polym. Sci. *257,* 43 (1979)
72. Koppelmann, J. et al.: Colloid Polym. Sci. *258,* 9 (1980)
73. Koppelmann, J.: Colloid Polym. Sci. *258,* 506 (1980)
74. Cuddihy, E. F., Moacanin, J.: J. Polym. Sci. A-2, *8,* 1627 (1970)
75. Yamashita, Y., Hikichi, K., Kaneko, M.: Polymer J. *12,* 439 (1980)
76. Holzmüller, W.: Adv. Polym. Sci. *26,* 1 (1978)
77. Schwarzl, F.: Kolloid Z. *165,* 88 (1959)
78. Gisolf, J. H.: Colloid Polym. Sci. *253,* 185 (1975)
79. Cowie, J. M. G.: Pure Appl. Chem. *51,* 2331 (1979)
80. Heijboer, J.: J. Polym. Sci. *C16,* 3413 (1968)
81. McCall, D. W.: J. Phys. Chem. *70,* 949 (1966)
82. Hayakawa, R., Tanabe, Y., Wada, Y.: Rep. Prog. Polym. Phys. Jpn. *15,* 301 (1972)
83. Hayakawa, R., Tanabe, Y., Wada, Y.: J. Macromol. Sci., Phys. *B8,* 445 (1973)
84. Hotta, Y., Tsuge, K., Wada, Y.: Rep. Prog. Polym. Phys. Jpn. *8,* 325 (1965)
85. Hayakawa, R., Wada, Y.: Res. Prog. Polym. Phys. Jpn. *16,* 347 (1973)
86. Hayakawa, R., Wada, Y.: J. Polym. Sci., Polym. Phys. Ed. *12,* 2119 (1974)
87. Tanabe, Y. et al.: Polym. J. *1,* 107 (1970)
88. Shimizu, K., Kawaruma, Y., Wada, Y.: Rep. Prog. Polym. Phys. Jpn. *15,* 305 (1972)
89. Sauer, J. A., Saba, R. G.: J. Macromol. Sci., Chem. *A3,* 1217 (1969)
90. Sauer, J. A.: J. Polym. Sci. *C32,* 69 (1971)
91. Reich, S., Eisenberg, A.: J. Polym. Sci. A-2, *10,* 1397 (1972)
92. Miller, S., Tomozawa, M., MacCrone, R. K.: Amorphous materials, Douglas, R. W., Ellis, B. (eds.). London: Wiley 1972
93. Papir, Y. S., Baer, E.: J. Appl. Phys. *42,* 4667 (1971)
94. Lee, S., Simha, R.: Macromolecules *7,* 909 (1974)
95. Stachurski, Z. H., Ward, I. M.: J. Polym. Sci. *A-2,* 6, 1817 (1968)
96. Illers, K. H.: Colloid Polym. Sci. *252,* 1 (1974)

97. Boyd, R. H., Breitling, S. M.: Polym. Prepr., Am. Chem. Soc. *14,* 192 (1973)
98. Boyd, R. H.: J. Polym. Sci., Polym. Phys. Ed. *13,*2345 (1975)
99. Boyd, R. H.: Polym. Eng. Sci. *16,* 365 (1976)
100. Kolařík, J., Janáček, J.: J. Polym. Sci. *C16,* 441 (1967)
101. Papir, Y. S., Kapur, S., Rogers, C. E., Baer, E.: J. Polym. Sci. *A-2,* 1305 (1972)
102. Baer, E., Kohn, R., Papir, Y. S.: J. Macromol. Sci., Phys. *B6,* 761 (1972)
103. Weber, G.: Angew. Makromol. Chem. *74,* 187 (1978)
104. Jacobs, H., Jenckel, E.: Makromol. Chem. *47,* 72 (1961)
105. Illers, K. H., Breuer, H.: J. Colloid Sci. *18,* 1 (1963)
106. Mocherla, K. K., Bell, J. P.: J. Polym. Sci., Polym. Phys. Ed. *11,* 1789 (1973)
107. Sacher, E.: J. Macromol. Sci., Phys. *B15,* 257 (1978)
108. Armeniades, C. D., Baer, E.: J. Polym. Sci. *A-2,* 9, 1345 (1971)
109. Kastelic, R., Baer, E.: J. acromol. Sci., Phys. *B7,* 679 (1973)
110. Pogany, G. A.: Polymer *11,* 66 (1970)
111. Wada, E., Fukuzawa, Y., Kurita, K.: Rep. Prog. Polym. Phys. Jpn. 18, 323 (1975)
112. Williams, J. G.: J. Appl. Polym. Sci. *23,* 3433 (1979)
113. Yamafuji, K., Ishida, Y.: Kolloid Z., Z. Polym. *183,*15 (1962)
114. Gisolf, J. H., Juijn, J. A.: Kolloid Z., Z. Polym. *239,* 545 (1970)
115. Harell, E. R., Chartoff, R. D.: Polymer Eng. Sci. *14,* 362 (1974)
116. Papir, Y. S., Baer, E.: Mater. Sci. Eng. *8,* 310 (1971)
117. Helfand, E.: J. Chem. Phys. *54,* 4651 (1971)
118. Skolnick, J., Helfand, E.: J. Chem. Phys. *72,* 5489 (1980)
119. Pechhold, W.: J. Polym. Sci. *C32,* 123 (1971)
120. Pechhold, W., Blasenbrey, S.: Angew. Makromol. Chem. *22,* 3 (1972)
121. Schatzki, T. F.: J. Polym. Sci. *57,* 496 (1962)
122. Schatzki, T. F.: Polym. Prepr., Am. Chem. Sco. *6,* 646 (1965)
123. Olf, H. G., Peterlin. A.: J. Polym. Sci. *A-2,* 8, 791 (1970)
124. Boyer, R. F.: Rubber Chem. Technol. *36,* 1303 (1963)
125. Wunderlich, B.: J. Chem. Phys. *37,* 2429 (1962)
126. Heijboer, J.: Physics of non-crystalline solids, Prins, J. A. (ed.), Amsterdam: North-Holland Publ. Co. 1965
127. Kolařík, J., Vacík, J., Kopeček, J.: Int. J. Polym. Mater. *3,* 259 (1975)
128. Cowie, J. M. G. et al.: Polymer *18,* 612 (1977)
129. Russell, G. A. et al.: J. Polym. Sci., Polym. Phys. Ed. *18,* 1271 (1980)
130. Ishida, Y., Yamafuji, K.: Kolloid Z. *177,* 97 (1961)
131. Williams, G., Cook, M.: Trans. Faraday Soc. *67,* 990 (1971)
132. Williams, G.: Adv. Polym. Sci. *33,* 60 (1979)
133. Cowie, J. M. G., McEwen, I. J., Velickovic, J.: Polymer *16,* 869 (1975)
134. Seefried, C. G., Koleske, J. V.: J. Polym. Sci., Polym. Phys. Ed. *14,* 663 (1976)
135. Gall, W. G., McCrum, N. G.: J. Polym. Sci. *50,* 489 (1961)
136. Kolařík, J., Dušek, K.: J. Macromol. Sci., Phys.*B10,* 157 (1974)
137. Eaton, R. F. et al.: Polym. Prepr., Am. Chem. Soc. *17,* 54 (1976)
138. Baer, E., Kohn, R., Papir, Y. S.: J. Macromol. Sci., Phys. *B6,* 761 (1972)
139. Allen, G., McAinsh, J., Jeffs, G. M.: Polymer *12,* 85 (1971)
140. Hiltner, A., Baer, E.: Polymer *15,* 805 (1974)
141. Vanderschueren, J., Linkens, A.: Macromolecules *11,* 1228 (1978)
142. Hains, P., Williams, G.: Polymer *16,* 715 (1976)
143. Puffr, R., Šebenda, J.: J. Polym. Sci. *C16,* 79 (1967)
144. Schroeder, L. R., Cooper, S. L.: J. Appl. Phys. *47,* 4310 (1976)
145. Pezzin, G.: Polym. Eng. Sci. *18,* 821 (1978)
146. Janáček, J., Kolařík, J.: Collect. Czech. Chem. Commun. *30,* 1597 (1965)
147. Janáček, J.: J. Polym. Sci. *C23,* 373 (1968)
148. Kolařík, J., Janáček, J.: J. Polym. Sci. *A-2,* 10, 11 (72)
149. Kolařík, J., Štol, M.: Polymer J. *5,* 158 (1973)
150. Kolařík, J., Janáček, J.: Collect. Czech. Chem.Commun. *32,* 1599 (1967)
151. Janáček, J., Kolařík, J.: J. Polym. Sci. *C16,* 279 (1967)
152. Kolařík, J.: J. Macromol. Sci., Phys. *B5,*355 (1971)

153. Kolařík, J.: Int. J. Polym. Mater. *1*, 125 (1972)
154. Janáček, J. et al.: Int. J. Polym. Mater. *5*,59 (1976)
155. Kolařík, J. et al.: Int. J. Polym. Mater. *5*, 89 (1976)
156. Kolařík, J.: J. Macromol. Sci. *B15*, 371 (1978)
157. Kolařík, J.: J. Appl. Polym. Sci. *24*, 1565 (1979)
158. Lochmann, L. et al.: J. Polym. Sci., Polym. Chem. Ed. *17*, 1727 (1979)
159. Kolařík, J.: Int. J. Polym. Mater. *8*, 275 (1980)
160. Kolařík, J., Murtinger, K., Ševčík, S.: to be published
161. Kolařík, J., Švantner, J., Janáček, J.: Chem. Papers *70*, 82 (1976) (in Czech)
162. Van Krevelen, D. W., Hoftyzer, P. J.: Properties of Polymers. Amsterdam: Elsevier 1972
163. MacKnight, W. J. et al.: J. Phys. Chem. *72*, 1122 (1968)
164. Razinskaya, I. N., Charitonova, N. E., Shtarkman, B. P.: Vysokomol. Soedin. *B11*, 892 (1969)
165. Boyer, R. F.: J. Polym. Sci. *C50*, 189 (1975)
166. Janáček, J.: J. Polym. Sci., Polym. Lett. Ed. *12*, 569 (1974)
167. Illers, K. H.: Ber. Bunsenges. Phys. Chem. *70*, 353 (1966)
168. Lednický, F., Janáček, J.: J. Macromol. Sci., Phys. *B5*, 335 (1971)
169. Bueche, F.: Physical properties of polymers. New York:Interscience 1962
170. Ilavský, M., Hasa, J.: Collect. Czech. Chem. Commun. *34*, 2199 (1969)
171. Rasmussen, D. H., MacKenzie, A. P.: J. Phys. Chem. *75*, 967 (1971)
172. Dušek, K., Bohdanecký, M., Prokopová, E.: Eur. Polym. J. *10*, 239 (1974)
173. Lange, N. A.: Handbook of Chemistry. New York: McGraw-Hill 1961
174. Radom, L. et al.: Aust. J. Chem. *25*, 1601 (1972)
175. Kolařík, J. et al.: J. Appl. Polym. Sci. *23*, 1553 (1979)
176. Manson, J. A., Sperling, L. H.: Polymer blends and composites. New York: Plenum Press 1976
177. Heijboer, J.: Brit. Polym. J. *1*, 3 (1969)
178. Locati, G., Tobolsky, A. V.: Advan. Mol. Relaxation Processes *1*, 375 (1970)
179. Vincent, P. I.: Polymer *15*, 111 (1974)
180. Carbuglio, C. et al.: J. Appl. Polym. Sci. *15*, 2487 (1971)
181. Andrews, E. H.: Fracture in polymers. London: Oliver & Boyd 1968
182. Manson, J. A. et al.: Polymer *16*, 850 (1975)
183. Hertzberg, R. W., Manson, J. A., Skibo, M. D.: Polymer *19*, 358 (1978)
184. Mascia, L.: Polymer *19*, 325 (1978)
185. Paul, D. R., Newman, S. (eds.): Polymer blends. New York: Academic Press 1978
186. Kolařík, J., Janáček, J., Nicolais, L.: J. Appl. Polym. Sci *20*, 841 (1976)
187. Takayanagi, M.: Pure Appl. Chem. *23*, 151 (1970)
188. Takayanagi, M.: J. Macromol. Sci., Phys. *B9*, 391 (1974)
189. Masui, R., Wada, Y.: J. Macromol. Sci., Phys. *B10*, 71 (1974)
190. MacKnight, W. J., Kajiyama, T., McKenna, L.: Polym. Eng. Sci. *8*, 267 (1968)
191. Tsutsui, T., Tanaka, R., Tanaka, T.: J. Polym. Sci., Polym. Phys. Ed. *14*, 2273 (1976)
192. Papanicolaou, G. C., Paipetis, S. A., Theocaris, P. S.: J. Appl. Polym. Sci. *20*, 903 (1976)
193. Yano, S.: J. Appl. Polym. Sci.*21*, 2645 (1977)
194. Yano, S., Murayama, M.: J. Appl. Polym. Sci. *25*, 433 (1980)

Received October 14, 1981
K. Dušek (editor)

Author Index Volumes 1–46

Allegra, G. and *Bassi, I. W.*: Isomorphism in Synthetic Macromolecular Systems. Vol. 6, pp. 549–574.

Andrews, E. H.: Molecular Fracture in Polymers. Vol. 27, pp. 1–66.

Anufrieva, E. V. and *Gotlib, Yu. Ya.*: Investigation of Polymers in Solution by Polarized Luminescence. Vol. 40, pp.1–68.

Arridge, R. C. and *Barham, P. J.*: Polymer Elasticity. Discrete and Continuum Models. Vol. 46, pp. 67–117.

Ayrey, G.: The Use of Isotopes in Polymer Analysis. Vol. 6, pp. 128–148.

Baldwin, R. L.: Sedimentation of High Polymers. Vol. 1, pp. 451–511.

Basedow, A, M. and *Ebert, K.*: Ultrasonic Degradation of Polymers in Solution. Vol. 22, pp. 83–148.

Batz, H.-G.: Polymeric Drugs. Vol. 23, pp. 25–53.

Bekturov, E. A. and *Bimendina, L. A.*: Interpolymer Complexes. Vol. 41, pp. 99–147.

Bergsma, F. and *Kruissink, Ch. A.*: Ion-Exchange Membranes. Vol 2, pp. 307–362.

Berlin, Al. Al., *Volfson, S. A.*, and *Enikolopian, N. S.*: Kinetics of Polymerization Processes. Vol. 38, pp. 89–140.

Berry, G. C. and *Fox, T. G.*: The Viscosity of Polymers and Their Concentrated Solutions. Vol. 5, pp. 261–357.

Bevington, J. C.: Isotopic Methods in Polymer Chemistry. Vol. 2, pp. 1–17.

Bird, R. B., *Warner, Jr., H. R.*, and *Evans, D. C.*: Kinetik Theory and Rheology of Dumbbell Suspensions with Brownian Motion. Vol. 8, pp. 1–90.

Biswas, M. and *Maity, C.*: Molecular Sieves as Polymerization Catalysts. Vol. 31, pp. 47–88.

Block, H.: The Nature and Application of Electrical Phenomena in Polymers. Vol. 33, pp. 93–167.

Böhm, L. L., *Chmeliř, M.*, *Löhr, G.*, *Schmitt, B. J.* und *Schulz, G. V.*: Zustände und Reaktionen des Carbanions bei der anionischen Polymerisation des Styrols. Vol. 9, pp. 1–45.

Bovey, F. A. and *Tiers, G. V. D.*: The High Resolution Nuclear Magnetic Resonance Spectroscopy of Polymers. Vol. 3, pp. 139–195.

Braun, J.-M. and *Guillet, J. E.*: Study of Polymers by Inverse Gas Chromatography. Vol. 21, pp. 107–145.

Breitenbach, J. W., *Olaj, O. F.* und *Sommer, F.*: Polymerisationsanregung durch Elektrolyse. Vol. 9, pp. 47–227.

Bresler, S. E. and *Kazbekov, E. N.*: Macroradical Reactivity Studied by Electron Spin Resonance. Vol. 3, pp. 688–711.

Bucknall, C. B.: Fracture and Failure of Multiphase Polymers and Polymer Composites. Vol. 27, pp. 121–148.

Bywater, S.: Polymerization Initiated by Lithium and Its Compounds. Vol. 4, pp. 66–110.

Bywater, S.: Preparation and Properties of Star-branched Polymers. Vol. 30, pp. 89–116.

Candau, S., *Bastide, J.* and *Delsanti, M.*: Structural, Elastic and Dynamic Properties of Swollen Polymer Networks. Vol. 44, pp. 27–72.

Carrick, W. L.: The Mechanism of Olefin Polymerization by Ziegler-Natta Catalysts. Vol. 12, pp. 65–86.

Casale, A. and *Porter, R. S.*: Mechanical Synthesis of Block and Graft Copolymers. Vol. 17, pp. 1–71.

Cerf, R.: La dynamique des solutions de macromolecules dans un champ de vitesses. Vol. 1, pp. 382–450.

Cesca, S., *Priola, A.* and *Bruzzone, M.*: Synthesis and Modification of Polymers Containing a System of Conjugated Double Bonds. Vol. 32, pp. 1–67.
Cicchetti, O.: Mechanisms of Oxidative Photodegradation and of UV Stabilization of Polyolefins. Vol. 7, pp. 70–112.
Clark, D. T.: ESCA Applied to Polymers. Vol. 24, pp. 125–188.
Coleman, Jr., L. E. and *Meinhardt, N. A.*: Polymerization Reactions of Vinyl Ketones. Vol. 1, pp. 159–179.
Crescenzi, V.: Some Recent Studies of Polyelectrolyte Solutions. Vol. 5, pp. 358–386.
Davydov, B. E. and *Krentsel, B. A.*: Progress in the Chemistry of Polyconjugated Systems. Vol. 25, pp. 1–46.
Dole, M.: Calorimetric Studies of States and Transitions in Solid High Polymers. Vol. 2, pp. 221–274.
Dreyfuss, P. and *Dreyfuss, M. P.*: Polytetrahydrofuran. Vol. 4, pp. 528–590.
Dušek, K. and *Prins, W.*: Structure and Elasticity of Non-Crystalline Polymer Networks. Vol. 6, pp. 1–102.
Eastham, A. M.: Some Aspects of the Polymerization of Cyclic Ethers. Vol. 2, pp. 18–50.
Ehrlich, P. and *Mortimer, G. A.*: Fundamentals of the Free-Radical Polymerization of Ethylene. Vol. 7, pp. 386–448.
Eisenberg, A.: Ionic Forces in Polymers. Vol. 5, pp. 59–112.
Elias, H.-G., *Bareiss, R.* und *Watterson, J. G.*: Mittelwerte des Molekulargewichts und anderer Eigenschaften. Vol. 11, pp. 111–204.
Elyashevich, G. K.: Thermodynamics and Kinetics of Orientational Crystallization of Flexible-Chain Polymers. Vol. 43, pp. 207–246.
Fischer, H.: Freie Radikale während der Polymerisation, nachgewiesen und identifiziert durch Elektronenspinresonanz. Vol. 5, pp. 463–530.
Fradet, A. and *Maréchal, E.*: Kinetics and Mechanisms of Polyesterifications. I. Reactions of Diols with Diacids. Vol. 43, pp. 51–144.
Fujita, H.: Diffusion in Polymer-Diluent Systems. Vol. 3, pp. 1–47.
Funke, W.: Über die Strukturaufklärung vernetzter Makromoleküle, insbesondere vernetzter Polyesterharze, mit chemischen Methoden. Vol. 4, pp. 157–235.
Gal'braikh, L. S. and *Rogovin, Z. A.*: Chemical Transformations of Cellulose. Vol. 14, pp. 87–130.
Gallot, B. R. M.: Preparation and Study of Block Copolymers with Ordered Structures, Vol. 29, pp. 85–156.
Gandini, A.: The Behaviour of Furan Derivatives in Polymerization Reactions. Vol. 25, pp. 47–96.
Gandini, A. and *Cheradame, H.*: Cationic Polymerization. Initiation with Alkenyl Monomers. Vol. 34/35, pp. 1–289.
Geckeler, K., *Pillai, V. N. R.*, and *Mutter, M.*: Applications of Soluble Polymeric Supports. Vol. 39, pp. 65–94.
Gerrens, H.: Kinetik der Emulsionspolymerisation. Vol. 1, pp. 234–328.
Ghiggino, K. P., *Roberts, A. J.* and *Phillips, D.*: Time-Resolved Fluorescence Techniques in Polymer and Biopolymer Studies. Vol. 40, pp. 69–167.
Goethals, E. J.: The Formation of Cyclic Oligomers in the Cationic Polymerization of Heterocycles. Vol. 23, pp. 103–130.
Graessley, W. W.: The Etanglement Concept in Polymer Rheology. Vol. 16, pp. 1–179.
Hagihara, N., *Sonogashira, K.* and *Takahashi, S.*: Linear Polymers Containing Transition Metals in the Main Chain. Vol. 41, pp. 149–179.
Hasegawa, M.: Four-Center Photopolymerization in the Crystalline State. Vol. 42, pp. 1–49.
Hay, A. S.: Aromatic Polyethers. Vol. 4, pp. 496–527.
Hayakawa, R. and *Wada, Y.*: Piezoelectricity and Related Properties of Polymer Films. Vol. 11, pp. 1–55.
Heidemann, E. and *Roth, W.*: Synthesis and Investigation of Collagen Model Peptides. Vol. 43, pp. 145–205.
Heitz, W.: Polymeric Reagents. Polymer Design, Scope, and Limitations. Vol. 23, pp. 1–23.
Helfferich, F.: Ionenaustausch. Vol. 1, pp. 329–381.
Hendra, P. J.: Laser-Raman Spectra of Polymers. Vol. 6, pp. 151–169.
Henrici-Olivé, G. und *Olivé, S.*: Kettenübertragung bei der radikalischen Polymerisation. Vol. 2, pp. 496–577.

Henrici-Olivé, G. und *Olivé, S.*: Koordinative Polymerisation an löslichen Übergangsmetall-Katalysatoren. Vol. 6, pp. 421–472.
Henrici-Olivé, G. and *Olivé, S.*: Oligomerization of Ethylene with Soluble Transition-Metal Catalysts. Vol. 15, pp. 1–30.
Henrici-Olivé, G. and *Olivé, S.*: Molecular Interactions and Macroscopic Properties of Polyacrylonitrile and Model Substances. Vol. 32, pp. 123–152.
Hermans, Jr., J., Lohr, D. and *Ferro, D.*: Treatment of the Folding and Unfolding of Protein Molecules in Solution According to a Lattic Model. Vol. 9, pp. 229–283.
Holzmüller, W.: Molecular Mobility, Deformation and Relaxation Processes in Polymers. Vol. 26, pp. 1–62.
Hutchison, J. and *Ledwith, A.*: Photoinitiation of Vinyl Polymerization by Aromatic Carbonyl Compounds. Vol. 14, pp. 49–86.
Iizuka, E.: Properties of Liquid Crystals of Polypeptides: with Stress on the Electromagnetic Orientation. Vol. 20, pp. 79–107.
Ikada, Y.: Characterization of Graft Copolymers. Vol. 29, pp. 47–84.
Imanishi, Y.: Syntheses, Conformation, and Reactions of Cyclic Peptides. Vol. 20, pp. 1–77.
Inagaki, H.: Polymer Separation and Characterization by Thin-Layer Chromatography. Vol. 24, pp. 189–237.
Inoue, S.: Asymmetric Reactions of Synthetic Polypeptides. Vol. 21, pp. 77–106.
Ise, N.: Polymerizations under an Electric Field. Vol. 6, pp. 347–376.
Ise, N.: The Mean Activity Coefficient of Polyelectrolytes in Aqueous Solutions and Its Related Properties. Vol. 7, pp. 536–593.
Isihara, A.: Intramolecular Statistics of a Flexible Chain Molecule. Vol. 7, pp. 449–476.
Isihara, A.: Irreversible Processes in Solutions of Chain Polymers. Vol. 5, pp. 531–567.
Isihara, A. and *Guth, E.*: Theory of Dilute Macromolecular Solutions. Vol. 5, pp. 233–260.
Janeschitz-Kriegl, H.: Flow Birefringence of Elastico-Viscous Polymer Systems. Vol. 6, pp. 170–318.
Jenkins, R. and *Porter, R. S.*: Unpertubed Dimensions of Stereoregular Polymers. Vol. 36, pp. 1–20.
Jenngins, B. R.: Electro-Optic Methods for Characterizing Macromolecules in Dilute Solution. Vol. 22, pp. 61–81.
Johnston, D. S.: Macrozwitterion Polymerization. Vol. 42, pp. 51–106.
Kamachi, M.: Influence of Solvent on Free Radical Polymerization of Vinyl Compounds. Vol. 38, pp. 55–87.
Kawabata, S. and *Kawai, H.*: Strain Energy Density Functions of Rubber Vulcanizates from Biaxial Extension. Vol. 24, pp. 89–124.
Kennedy, J. P. and *Chou, T.*: Poly(isobutylene-*co*-β-Pinene): A New Sulfur Vulcanizable, Ozone Resistant Elastomer by Cationic Isomerization Copolymerization. Vol. 21, pp. 1–39.
Kennedy, J. P. and *Delvaux, J. M.*: Synthesis, Characterization and Morphology of Poly(butadiene-*g*-Styrene). Vol. 38, pp. 141–163.
Kennedy, J. P. and *Gillham, J. K.*: Cationic Polymerization of Olefins with Alkylaluminium Initiators. Vol. 10, pp. 1–33.
Kennedy, J. P. and *Johnston, J. E.*: The Cationic Isomerization Polymerization of 3-Methyl-1-butene and 4-Methyl-1-pentene. Vol. 19, pp. 57–95.
Kennedy, J. P. and *Langer, Jr., A. W.*: Recent Advances in Cationic Polymerization. Vol. 3, pp. 508–580.
Kennedy, J. P. and *Otsu, T.*: Polymerization with Isomerization of Monomer Preceding Propagation. Vol. 7, pp. 369–385.
Kennedy, J. P. and *Rengachary, S.*: Correlation Between Cationic Model and Polymerization Reactions of Olefins. Vol. 14, pp. 1–48.
Kennedy, J. P. and *Trivedi, P. D.*: Cationic Olefin Polymerization Using Alkyl Halide – Alkylaluminum Initiator Systems. I. Reactivity Studies. II. Molecular Weight Studies. Vol. 28, pp. 83–151.
Kennedy, J. P., Chang, V. S. C. and *Guyot, A.*: Carbocationic Synthesis and Characterization of Polyolefins with Si–H and Si–Cl Head Groups. Vol. 43, pp. 1–50.
Khoklov, A. R. and *Grosberg, A. Yu.*: Statistical Theory of Polymeric Lyotropic Liquid Crystals. Vol. 41, pp. 53–97.

Kissin, Yu. V.: Structures of Copolymers of High Olefins. Vol. 15, pp. 91–155.
Kitagawa, T. and *Miyazawa, T.*: Neutron Scattering and Normal Vibrations of Polymers. Vol. 9, pp. 335–414.
Kitamaru, R. and *Horii, F.*: NMR Approach to the Phase Structure of Linear Polyethylene. Vol. 26., pp. 139–180.
Knappe, W.: Wärmeleitung in Polymeren. Vol. 7, pp. 477–535.
Kolařík, J.: Secondary Relaxations in Glassy Polymers: Hydrophylic Polymethacrylates and Polyacrylates: Vol. 46, pp. 119–161.
Koningsveld, R.: Preparative and Analytical Aspects of Polymer Fractionation. Vol. 7.
Kovacs, A. J.: Transition vitreuse dans les polymers amorphes. Etude phénoménologique. Vol. 3, pp. 394–507.
Krässig, H. A.: Graft Co-Polymerization of Cellulose and Its Derivatives. Vol. 4, pp. 111–156.
Kraus, G.: Reinforcement of Elastomers by Carbon Black. Vol. 8, pp. 155–237.
Kreutz, W. and *Welte, W.*: A General Theory for the Evaluation of X-Ray Diagrams of Biomembranes and Other Lamellar Systems. Vol. 30, pp. 161–225.
Krimm, S.: Infrared Spectra of High Polymers. Vol. 2, pp. 51–72.
Kuhn, W., *Ramel, A.*, *Walters, D. H.*, *Ebner, G.* and *Kuhn, H. J.*: The Production of Mechanical Energy from Different Forms of Chemical Energy with Homogeneous and Cross-Striated High Polymer Systems. Vol. 1, pp. 540–592.
Kunitake, T. and *Okahata, Y.*: Catalytic Hydrolysis by Synthetic Polymers. Vol. 20, pp. 159–221.
Kurata, M. and *Stockmayer, W. H.*: Intrinsic Viscosities and Unperturbed Dimensions of Long Chain Molecules. Vol. 3, pp. 196–312.
Ledwith, A. and *Sherrington, D. C.*: Stable Organic Cation Salts: Ion Pair Equilibria and Use in Cationic Polymerization. Vol. 19, pp. 1–56.
Lee, C.-D. S. and *Daly, W. H.*: Mercaptan-Containing Polymers. Vol. 15, pp. 61–90.
Lipatov, Y. S.: Relaxation and Viscoelastic Properties of Heterogeneous Polymeric Compositions. Vol. 22, pp. 1–59.
Lipatov, Y. S.: The Iso-Free-Volume State and Glass Transitions in Amorphous Polymers: New Development of the Theory. Vol. 26, pp. 63–104.
Mano, E. B. and *Coutinho, F. M. B.*: Grafting on Polyamides. Vol. 19, pp. 97–116.
Mark, J. E.: The Use of Model Polymer Networks to Elucidate Molecular Aspects of Rubberlike Elasticity. Vol. 44, pp. 1–26.
Mengoli, G.: Feasibility of Polymer Film Coating Through Electroinitiated Polymerization in Aqueous Medium. Vol. 33, pp. 1–31.
Meyerhoff, G.: Die viscosimetrische Molekulargewichtsbestimmung von Polymeren. Vol. 3, pp. 59–105.
Millich, F.: Rigid Rods and the Characterization of Polyisocyanides. Vol. 19, pp. 117–141.
Morawetz, H.: Specific Ion Binding by Polyelectrolytes. Vol. 1, pp. 1–34.
Morin, B. P., *Breusova, I. P.* and *Rogovin, Z. A.*: Structural and Chemical Modifications of Cellulose by Graft Copolymerization. Vol. 42, pp. 139–166.
Mulvaney, J. E., *Oversberger, C. C.* and *Schiller, A. M.*: Anionic Polymerization. Vol. 3, pp. 106–138.
Okubo, T. and *Ise, N.*: Synthetic Polyelectrolytes as Models of Nucleic Acids and Esterases. Vol. 25, pp. 135–181.
Osaki, K.: Viscoelastic Properties of Dilute Polymer Solutions. Vol. 12, pp. 1–64.
Oster, G. and *Nishijima, Y.*: Fluorescence Methods in Polymer Science. Vol. 3, pp. 313–331.
Overberger, C. G. and *Moore, J. A.*: Ladder Polymers. Vol. 7, pp. 113–150.
Patat, F., *Killmann, E.* und *Schiebener, C.*: Die Absorption von Makromolekülen aus Lösung. Vol. 3, pp. 332–393.
Penczek, S., *Kubisa, P.* and *Matyjaszewski, K.*: Cationic Ring-Opening Polymerization of Heterocyclic Monomers. Vol. 37, pp. 1–149.
Peticolas, W. L.: Inelastic Laser Light Scattering from Biological and Synthetic Polymers. Vol. 9, pp. 285–333.
Pino, P.: Optically Active Addition Polymers. Vol. 4, pp. 393–456.
Plate, N. A. and *Noah, O. V.*: A Theoretical Consideration of the Kinetics and Statistics of Reactions of Functional Groups of Macromolecules. Vol. 31, pp. 133–173.
Plesch, P. H.: The Propagation Rate-Constants in Cationic Polymerisations. Vol. 8, pp. 137–154.

Porod, G.: Anwendung und Ergebnisse der Röntgenkleinwinkelstreuung in festen Hochpolymeren. Vol. 2, pp. 363–400.
Pospíšil, J.: Transformations of Phenolic Antioxidants and the Role of Their Products in the Long-Term Properties of Polyolefins. Vol. 36, pp. 69–133.
Postelnek, W., *Coleman, L. E.*, and *Lovelace, A. M.*: Fluorine-Containing Polymers. I. Fluorinated Vinyl Polymers with Functional Groups, Condensation Polymers, and Styrene Polymers. Vol. 1, pp. 75–113.
Rempp, P., *Herz, J.*, and *Borchard, W.*: Model Networks. Vol. 26, pp. 107–137.
Rigbi, Z.: Reinforcement of Rubber by Carbon Black. Vol. 36, pp. 21–68.
Rogovin, Z. A. and *Gabrielyan, G. A.*: Chemical Modifications of Fibre Forming Polymers and Copolymers of Acrylonitrile. Vol. 25, pp. 97–134.
Roha, M.: Ionic Factors in Steric Control. Vol. 4, pp. 353–392.
Roha, M.: The Chemistry of Coordinate Polymerization of Dienes. Vol. 1, pp. 512–539.
Safford, G. J. and *Naumann, A. W.*: Low Frequency Motions in Polymers as Measured by Neutron Inelastic Scattering. Vol. 5, pp. 1–27.
Schuerch, C.: The Chemical Synthesis and Properties of Polysaccharides of Biomedical Interest. Vol. 10, pp. 173–194.
Schulz, R. C. und *Kaiser, E.*: Synthese und Eigenschaften von optisch aktiven Polymeren. Vol. 4, pp. 236–315.
Seanor, D. A.: Charge Transfer in Polymers. Vol. 4, pp. 317–352.
Seidl, J., *Malinský, J.*, *Dušek, K.* und *Heitz, W.*: Makroporöse Styrol-Divinylbenzol-Copolymere und ihre Verwendung in der Chromatographie und zur Darstellung von Ionenaustauschern. Vol. 5, pp. 113–213.
Semjonow, V.: Schmelzviskositäten hochpolymerer Stoffe. Vol. 5, pp. 387–450.
Semlyen, J. A.: Ring-Chain Equilibria and the Conformations of Polymer Chains. Vol. 21, pp. 41–75.
Sharkey, W. H.: Polymerizations Through the Carbon-Sulphur Double Bond. Vol. 17, pp. 73–103.
Shimidzu, T.: Cooperative Actions in the Nucleophile-Containing Polymers. Vol. 23, pp. 55–102.
Shutov, F. A.: Foamed Polymers Based on Reactive Oligomers, Vol. 39, pp. 1–64.
Silvestri, G., *Gambino, S.*, and *Filardo, G.*: Electrochemical Production of Initiators for Polymerization Processes. Vol. 38, pp. 27–54.
Slichter, W. P.: The Study of High Polymers by Nuclear Magnetic Resonance. Vol. 1, pp. 35–74.
Small, P. A.: Long-Chain Branching in Polymers. Vol. 18.
Smets, G.: Block and Graft Copolymers. Vol. 2, pp. 173–220.
Sohma, J. and *Sakaguchi, M.*: ESR Studies on Polymer Radicals Produced by Mechanical Destruction and Their Reactivity. Vol. 20, pp. 109–158.
Sotobayashi, H. und *Springer, J.*: Oligomere in verdünnten Lösungen. Vol. 6, pp. 473–548.
Sperati, C. A. and *Starkweather, Jr., H. W.*: Fluorine-Containing Polymers. II. Polytetrafluoroethylene. Vol. 2, pp. 465–495.
Sprung, M. M.: Recent Progress in Silicone Chemistry. I. Hydrolysis of Reactive Silane Intermediates. Vol. 2, pp. 442–464.
Stahl, E. and *Brüderle, V.*: Polymer Analysis by Thermofractography. Vol. 30, pp. 1–88.
Stannett, V. T., *Koros, W. J.*, *Paul, D. R.*, *Lonsdale, H. K.*, and *Baker, R. W.*: Recent Advances in Membrane Science and Technology. Vol. 32, pp. 69–121.
Staverman, A. J.: Properties of Phantom Networks and Real Networks. Vol. 44, pp. 73–102.
Stauffer, D., *Coniglio, A.* and *Adam, M.*: Gelation and Critical Phenomena. Vol. 44, pp. 103–158.
Stille, J. K.: Diels-Alder Polymerization. Vol. 3, pp. 48–58.
Stolka, M. and *Pai, D.*: Polymers with Photoconductive Properties. Vol. 29, pp. 1–45.
Subramanian, R. V.: Electroinitiated Polymerization on Electrodes. Vol. 33, pp. 33–58.
Sumitomo, H. and *Okada, M.*: Ring-Opening Polymerization of Bicyclic Acetals, Oxalactone, and Oxalactam. Vol. 28, pp. 47–82.
Szegö, L.: Modified Polyethylene Terephthalate Fibers. Vol. 31, pp. 89–131.
Szwarc, M.: Termination of Anionic Polymerization. Vol. 2, pp. 275–306.
Szwarc, M.: The Kinetics and Mechanism of N-carboxy-α-amino-acid Anhydride (NCA) Polymerization to Poly-amino Acids. Vol. 4, pp. 1–65.
Szwarc, M.: Thermodynamics of Polymerization with Special Emphasis on Living Polymers. Vol. 4, pp. 457–495.

Takahashi, A. and *Kawaguchi, M.*: The Structure of Macromolecules Adsorbed on Interfaces. Vol. 46, pp. 1–65.
Takemoto, K. and *Inaki, Y.*: Synthetic Nucleic Acid Analogs. Preparation and Interactions. Vol. 41, pp. 1–51.
Tani, H.: Stereospecific Polymerization of Aldehydes and Epoxides. Vol. 11, pp. 57–110.
Tate, B. E.: Polymerization of Itaconic Acid and Derivatives. Vol. 5, pp. 214–232.
Tazuke, S.: Photosensitized Charge Transfer Polymerization. Vol. 6, pp. 321–346.
Teramoto, A. and *Fujita, H.*: Conformation-dependent Properties of Synthetic Polypeptides in the Helix-Coil Transition Region. Vol. 18, pp. 65–149.
Thomas, W. M.: Mechanism of Acrylonitrile Polymerization. Vol. 2, pp. 401–441.
Tobolsky, A. V. and *DuPré, D. B.*: Macromolecular Relaxation in the Damped Torsional Oscillator and Statistical Segment Models. Vol. 6, pp. 103–127.
Tosi, C. and *Ciampelli, F.*: Applications of Infrared Spectroscopy to Ethylene-Propylene Copolymers. Vol. 12, pp. 87–130.
Tosi, C.: Sequence Distribution in Copolymers: Numerical Tables. Vol. 5, pp. 451–462.
Tsuchida, E. and *Nishide, H.*: Polymer-Metal Complexes and Their Catalytic Activity. Vol. 24, pp. 1–87.
Tsuji, K.: ESR Study of Photodegradation of Polymers. Vol. 12, pp. 131–190.
Tsvetkov, V. and *Andreeva, L.*: Flow and Electric Birefringence in Rigid-Chain Polymer Solutions. Vol. 39, pp. 95–207.
Tuzar, Z., Kratochvíl, P., and *Bohdanecký, M.*: Dilute Solution Properties of Aliphatic Polyamides. Vol. 30, pp. 117–159.
Valvassori, A. and *Sartori, G.*: Present Status of the Multicomponent Copolymerization Theory. Vol. 5, pp. 28–58.
Voorn, M. J.: Phase Separation in Polymer Solutions. Vol. 1, pp. 192–233.
Werber, F. X.: Polymerization of Olefins on Supported Catalysts. Vol. 1, pp. 180–191.
Wichterle, O., Šebenda, J., and *Králiček, J.*: The Anionic Polymerization of Caprolactam. Vol. 2, pp. 578–595.
Wilkes, G. L.: The Measurement of Molecular Orientation in Polymeric Solids. Vol. 8, pp. 91–136.
Williams, G.: Molecular Aspects of Multiple Dielectric Relaxation Processes in Solid Polymers. Vol. 33, pp. 59–92.
Williams, J. G.: Applications of Linear Fracture Mechanics. Vol. 27, pp. 67–120.
Wöhrle, D.: Polymere aus Nitrilen. Vol. 10, pp. 35–107.
Wolf, B. A.: Zur Thermodynamik der enthalpisch und der entropisch bedingten Entmischung von Polymerlösungen. Vol. 10, pp. 109–171.
Woodward, A. E. and *Sauer, J. A.*: The Dynamic Mechanical Properties of High Polymers at Low Temperatures. Vol. 1, pp. 114–158.
Wunderlich, B. and *Baur, H.*: Heat Capacities of Linear High Polymers. Vol. 7, pp. 151–368.
Wunderlich, B.: Crystallization During Polymerization. Vol. 5, pp. 568–619.
Wrasidlo, W.: Thermal Analysis of Polymers. Vol. 13, pp. 1–99.
Yamashita, Y.: Random and Black Copolymers by Ring-Opening Polymerization. Vol. 28, pp. 1–46.
Yamazaki, N.: Electrolytically Initiated Polymerization. Vol. 6, pp. 377–400.
Yamazaki, N. and *Higashi, F.*: New Condensation Polymerizations by Means of Phosphorus Compounds. Vol. 38, pp. 1–25.
Yokoyama, Y. and *Hall H. K.*: Ring-Opening Polymerization of Atom-Bridged and Bond-Bridged Bicyclic Ethers, Acetals and Orthoesters. Vol. 42, pp. 107–138.
Yoshida, H. and *Hayashi, K.*: Initiation Process of Radiation-induced Ionic Polymerization as Studied by Electron Spin Resonance. Vol. 6, pp. 401–420.
Yuki, H. and *Hatada, K.*: Stereospecific Polymerization of Alpha-Substituted Acrylic Acid Esters. Vol. 31, pp. 1–45.
Zachmann, H. G.: Das Kristallisations- und Schmelzverhalten hochpolymerer Stoffe. Vol. 3, pp. 581–687.
Zambelli, A. and *Tosi, C.*: Stereochemistry of Propylene Polymerization. Vol. 15, pp. 31–60.

GPSR Compliance

The European Union's (EU) General Product Safety Regulation (GPSR) is a set of rules that requires consumer products to be safe and our obligations to ensure this.

If you have any concerns about our products, you can contact us on ProductSafety@springernature.com

In case Publisher is established outside the EU, the EU authorized representative is:

Springer Nature Customer Service Center GmbH
Europaplatz 3
69115 Heidelberg, Germany

Batch number: 10620881

Printed by Printforce, the Netherlands